List of titles

AF860676

Already published

Cell Differentiation	J. M. Ash
Biochemical Genetics	R. A. Woo
Functions of Biological Membranes	M. Davies
Cellular Development	D. Garrod
Brain Biochemistry	H. S. Bachelard
Immunochemistry	M. W. Steward
The Selectivity of Drugs	A. Albert
Biomechanics	R. McN. Alexander
Molecular Virology	T. H. Pennington, D. A. Ritchie
Hormone Action	A. Malkinson
Cellular Recognition	M. F. Greaves
Cytogenetics of Man and other Animals	A. McDermott
RNA Biosynthesis	R. H. Burdon
Protein Biosynthesis	A. E. Smith
Biological Energy Conservation	C. Jones
Control of Enzyme Activity	P. Cohen
Metabolic Regulation	R. Denton, C. I. Pogson
Plant Cytogenetics	D. M. Moore
Population Genetics	L. M. Cook
Insect Biochemistry	H. H. Rees
A Biochemical Approach to Nutrition	R. A. Freedland, S. Briggs
Enzyme Kinetics	P. C. Engel

In preparation

The Cell Cycle	S. Shall
Polysaccharide Shapes	D. A. Rees
Microbial Metabolism	H. Dalton, R. R. Eady
Bacterial Taxonomy	D. Jones, M. Goodfellow
Molecular Evolution	W. Fitch
Metal Ions in Biology	P. M. Harrison, R. Hoare
Cellular Immunology	D. Katz
Muscle	R. M. Simmons
Xenobiotics	D. V. Parke
Human Genetics	J. H. Edwards
Biochemical Systematics	J. B. Harborne
Biochemical Pharmacology	B. A. Callingham
Biological Oscillations	A. Robertson
Photobiology	K. Poff
Functional Aspects of Neurochemistry	G. B. Ansell, S. Spanner
Cellular Degradative Processes	R. J. Dean
Transport Phenomena in Plants	D. A. Baker
Membrane Assembly	J. Haslam

OUTLINE STUDIES IN BIOLOGY

Editor's Foreword

The student of biological science in his final years as an undergraduate and his first years as a graduate is expected to gain some familiarity with current research at the frontiers of his discipline. New research work is published in a perplexing diversity of publications and is inevitably concerned with the minutiae of the subject. The sheer number of research journals and papers also causes confusion and difficulties of assimilation. Review articles usually presuppose a background knowledge of the field and are inevitably rather restricted in scope. There is thus a need for short but authoritative introductions to those areas of modern biological research which are either not dealt with in standard introductory textbooks or are not dealt with in sufficient detail to enable the student to go on from them to read scholarly reviews with profit. This series of books is designed to satisfy this need. The authors have been asked to produce a brief outline of their subject assuming that their readers will have read and remembered much of a standard introductory textbook of biology. This outline then sets out to provide by building on this basis, the conceptual framework within which modern research work is progressing and aims to give the reader an indication of the problems, both conceptual and practical, which must be overcome if progress is to be maintained. We hope that students will go on to read the more detailed reviews and articles to which reference is made with a greater insight and understanding of how they fit into the overall scheme of modern research effort and may thus be helped to choose where to make their own contribution to this effort.

These books are guidebooks, not textbooks. Modern research pays scant regard for the academic divisions into which biological teaching and introductory textbooks must, to a certain extent, be divided. We have thus concentrated in this series on providing guides to those areas which fall between, or which involve, several different academic disciplines. It is here that the gap between the textbook and the research paper is widest and where the need for guidance is greatest. In so doing we hope to have extended or supplemented but not supplanted main texts, and to have given students assistance in seeing how modern biological research is progressing, while at the same time providing a foundation for self help in the achievement of successful examination results.

J. M. Ashworth, Professor of Biology, University of Essex.

Insect Biochemistry

H.H. Rees

Lecturer in Biochemistry,
University of Liverpool

LONDON
CHAPMAN AND HALL

A Halsted Press Book
JOHN WILEY & SONS, NEW YORK

First published in 1977
by Chapman and Hall Ltd.,
11 New Fetter Lane, London EC4P 4EE
© 1977 H.H. Rees
Typeset by Preface Ltd., Salisbury, Wilts and printed in Great Britain at the University Printing House, Cambridge

ISBN 0 412 13130 7

This paperback edition is sold subject to the condition that it shall not, by way of trade or otherwise, be lent, re-sold, hired out, or otherwise circulated without the publisher's prior consent in any form of binding or cover other than that in which it is published and without a similar condition including this condition being imposed on the subsequent purchaser.

All rights reserved. No part of this book may be reprinted, or reproduced or utilized in any form or by any electronic, mechanical or other means, now known or hereafter invented, including photocopying and recording, or in any information storage and retrieval system, without permission in writing from the Publisher.

Distributed in the U.S.A.
by Halsted Press, a Division
of John Wiley & Sons, Inc., New York

Library of Congress Cataloging in Publication Data

Rees, H. H.
Insect biochemistry.

1. Insects–Physiology. I. Title. [DNIM: 1. Biochemistry.
2. Insects. QX500 R328i]
QL495.R43 1977 595.7'01 76–46326
ISBN 0-470-98987-4

Contents

Acknowledgement

I am especially grateful to Drs G. Britton and E. Hilary Evans and Professor T. W. Goodwin, C.B.E., F.R.S., for their helpful comments and advice on the manuscript, and Mrs M. R. C. O'Neill for transcribing my handwriting into typescript.

1 Introduction

The success of the Insecta as a class (nearly 1 million spp.; phylum Arthropoda) is largely due to their adaptability to profoundly different ecological niches. Insects have attracted the attention of scientists both as useful model systems for the study of many basic biological phenomena, and also for the rational development of new methods of controlling the pest species. As a class, insects have played an important role in the elucidation of numerous basic biochemical phenomena. For example, work on the genetic control of eye pigment formation in insects led to the first clear demonstration of the basic one gene – one protein principle, and the initial work on the cytochromes which led to our present concepts of oxidative metabolism was carried out on insect flight muscle.

A virtue of insects as biological research material lies in their diversity. It is often possible to find a species where a particular biochemical process is greatly emphasized. However, this remarkable diversity can also be a serious complication in the development of a coherent, overall picture of biochemical processes within this class. It is dangerous to extrapolate results from one species to another, and to make generalizations.

The higher insects are particularly suitable subjects for morphogenetic studies in view of the existence of several morphologically well-defined stages. The metamorphosis (*see* p.8) of such insects represents advanced polymorphism, since the development of a fully grown, differentiated free-living organism (larva), which is extensively destroyed and rebuilt into a morphologically different form (adult) suitable for life in a different ecological niche, is controlled by a single genome. This is probably the most dramatic reorganization of a growing animal known. Certain carefully selected insect material can, thus, provide suitable model systems for developmental studies.

The majority of the individual metabolic reactions occurring in insects are similar to those found in other groups of organisms. However, the way that these reactions are integrated and regulated, particularly in individual tissues, often gives rise to distinctive systems. Insect flight muscle is of special interest as a model system, because oxidative metabolism there is greatly emphasized and subject to considerable regulation.

In this short treatment of insect biochemistry, emphasis will be placed upon the distinctive features, and, wherever feasible, will indicate why these have evolved to cope with differing ecological needs. An attempt will also be made to illustrate why insects are such useful tools for biochemical research, particularly in development and oxidative metabolism. Initially, however, a few aspects of growth and development require brief consideration.

1.1 Growth and development

Since the general body covering (integument) in insects also forms the exoskeleton, which is only capable of limited expansion, growth in these animals is intimately associated with moulting. The intervals between the moults are known as *stages* or *stadia,* and the form assumed by an insect during a particular stadium is termed an *instar.*

Most insects are hatched in a morphological condition which differs from that assumed in the adult (*imago*). Consequently, they have to pass through dramatic changes of form, which collectively are termed *metamorphosis.* During larval moults, the new cuticle is basically similar to the old one, and the internal organs undergo comparatively little change. The winged insect orders (Pterygota) can be divided into two groups based on the degree of metamorphosis which occurs: the *hemimetabola* and *holometabola.*

In the hemimetabola, the immature insects resemble the adults in several respects, but they lack wings, gonads, and external genitalia. The orders which belong to this group are the Exopterygota, and the developing wings are seen externally as 'wing pads'. The immature instars in this primitive group of insects are commonly referred to as *nymphs.* Although there is a gradual transition to the winged, reproducing adult, pronounced metamorphosis occurs at the nymphal-adult moult, when cuticle of a different nature is deposited with extensive histolysis of obsolete internal organs and growth of new ones suited for adult life e.g. flight muscles and gonads.

Holometabolous development is shown by the more highly evolved Endopterygota, in which wings develop internally. In this case, metamorphosis comprises two special moults, and the transformation of larva into adult occurs via a pupal instar, in which the larval tissues are largely destroyed and rebuilt into adult structures.

2 Distinctive features of metabolism and homeostasis

This section will mainly consider aspects of insect metabolism which are to some extent specialized or distinctive, and will relate these, wherever possible, to the particular needs of the organism. An important organ in insect metabolism is the fat body, which generally consists of an arrangement of lobes or sheets, allowing maximum area of contact with the haemolymph, and is thus admirably suited for rapid exchange of metabolites. The haemolymph in insects is the only extracellular fluid, which circulates through a poorly defined system of sinuses (haemocoele) and thus bathes all the organs. Although fat, glycogen and protein accumulate in the fat body during postembryonic feeding, it is also the site of considerable metabolic activity, and may, indeed, perform many of the functions associated with the mammalian liver.

2.1 Digestion and absorption

On the whole, the digestive enzymes present and their relative activities in a particular insect species generally reflect the composition of its normal diet. Some insect species (mainly larvae of wood-boring beetles) together with a few members of the Mollusca, are, apparently, the only animals to possess cellulase enzymes; cellulose digestion frequently involves symbiotic microorganisms [1,2].

The distinctive nature of insect metabolism has imposed corresponding specialization in the absorptive mechanisms of digested foodstuffs from the alimentary tract; these mechanisms also show appreciable species variation. For a critical discussion of gut absorption see [3].

In contrast to mammals, there is no evidence for active transport of monosaccharides through the gut wall in the few investigated insect species. Monosaccharides are apparently absorbed by passive diffusion, facilitated by their rapid conversion into haemolymph trehalose (α-D-glucopyranosyl-α-D-glucopyranoside), a process carried out by fat body investing the midgut. The doubling in molecular volume upon trehalose formation may help restrict back-diffusion through the pervious midgut. Since high trehalose and low glucose concentrations in the haemolymph do not universally occur in insects, mechanisms of sugar absorption besides diffusion may exist.

In the locust, *Schistocerca gregaria,* diffusion of amino acids from the midgut into the haemolymph occurs. This process is apparently linked to the withdrawal of water from the gut lumen at a more rapid rate than the net removal of amino acids, thus concentrating materials in the lumen and favouring the lumen/haemolymph diffusion gradient. Although differential absorption of amino acids has been observed in a number of insects, there is no evidence as yet for an active transport mechanism of amino acid uptake against concentration gradients as in vertebrates. It is possible that the uptake from the gut by diffusion may be correlated with the exceptionally high concentrations of amino acids in insect haemolymph, since an active transport mechanism would be energetically expensive in view of the extreme concentration gradients across the midgut wall.

The exact manner of lipid translocation across the gut wall in insects is uncertain.

The mechanisms of absorption of water and inorganic ions from the gut differ appreciably amongst the few species of insects investigated, but due to the paucity of information no clear patterns have emerged. Amongst these mechanistic divergencies a few striking specializations occur. For example, in the midgut of *Hyalophora cercropia*, active uptake of K^+ into the haemolymph occurs, but this ion pump is different from those identified in many vertebrate and invertebrate cells, since K^+ transport may be linked to movement of H^+ rather than involve equivalent exchange with Na^+. The presence of this specialized ion pump may be correlated with the unusual ionic composition of the diet in this species, since the ratio of K^+/Na^+ in the midgut fluid is particularly high. A somewhat analogous situation exists in certain marine organisms living in a freshwater environment when Na^+ uptake is linked to H^+ exchange. The mechanism of water uptake from the rectum of *S. gregaria* and *Calliphora* into the haemolymph is specialized in that the active transport of the water molecules is apparently independent of the net movements of inorganic ions across the rectal wall.

2.2 The flight muscle

2.2.1 Respiratory exchange

Insects have an elaborate tracheal respiratory system, which communicates directly with the exterior; the tracheae divide into fine capillaries (tracheoles), which always terminate in, or on, the surface of the cells of tissues. Thus, oxygen is normally conveyed directly to the cells without the participation of the haemolymph.

The overall level of metabolism in the working flight muscle is particularly intense, and the rate of oxygen uptake during flight can be as much as 50–100 times greater than the resting values [4]. For example, in the blowfly, *Lucilia sericata*, some individuals have an oxygen consumption in the region of 3000 μl/min/g upon initiation of flight. This is one of the most intense respiration rates so far reported in nature and also is the most highly controlled. In contrast, the flying humming bird, which is reputed to have the greatest respiratory rate per unit weight recorded amongst the vertebrates, exhibits a mere fivefold control. Insect flight muscle is thus clearly the tissue of choice as a model system for studies on the control of catabolism and biological oxidations [5,6].

Two morphologically distinguishable types of insect flight muscle occur, which have different mechanisms of wing oscillation. The Orthoptera and Lepidoptera, have the classical synchronous type of excitation–contraction coupling (afibrillar muscle morphology) and have relatively slow rates of wing beat frequency (5–30/s), which are dependent upon the activity of the central nervous system. A short burst of motor nerve impulse triggers each cycle of mechanical activity. The Diptera and Hymenoptera have the asynchronous type of excitation–contraction coupling (fibrillar muscle morphology) and are characterized by a high wingbeat frequency (often over 1000/s). In this case mechanical activity is sustained by a self-oscillatory mechanism, which is merely initiated and terminated by motor nerve control.

Insects can generally fly for some hours with enormous respiratory rates without developing a significant 'oxygen debt', indicating that their metabolism is not limited by oxygen availability. This is in marked contrast to mammalian metabolism, where an elevated respiratory rate generally persists for some time after a period of exercise to recover the 'oxygen debt'. Myoglobin and haemoglobin are absent from insect flight muscle, and air is conveyed directly to the tissue via the elaborate network of tracheae and tracheoles, which invade the fibres and end in close juxtaposition with the mitochondria. By exploiting the high atmospheric oxygen supply in this way, insect flight muscle has become totally committed to aerobic metabolism: contraction for even short periods of anoxia is not possible [7].

2.2.2 Nature of the substrate consumed during flight

In general, insects possessing the asynchronous type of flight muscle, during flight, have a respiratory quotient (RQ) approaching unity, carbohydrates being almost sole substrate [4]. On the other hand, in insects possessing the synchronous type of flight muscle, the RQ is in the region of 0·73 and fats become depleted. Locusts, cockroaches and aphids form an intermediate group of insects, using carbohydrates (glycogen and trehalose) during the initial period of flight, until these substrates essentially become depleted, when the RQ decreases as fat becomes the main fuel; this allows flight to be sustained for some hours. In some insects amino acids are important in flight muscle metabolism. Proline, for example, is rapidly utilized apparently to prime the citric acid cycle on initiation of flight in the blowfly, and is the principal energy reserve in the tsetse-fly (see pp. 17 and 18).

Many insects which oxidize fat either exclusively or principally during flight undertake migratory flights of long duration, although this ability to oxidize fat is not restricted to such migratory species. Fat has advantages as a metabolic fuel for continuous work in certain circumstances for many animals, e.g. the ATP yield per carbon unit is greater during oxidation of fatty acid than of glucose.

A sufficiently rapid rate of substrate supply to the mitochondria is of paramount importance for the maintenance of the rapid oxidative processes within the flight muscle. Since transport of haemolymph substrate into the interior of large flight muscles occurs by a combination of diffusion and muscular pumping, an inefficient mechanism, the haemolymph fuel concentration must be particularly high to provide concentration gradients for rapid supply of substrate to the interior of the fibres.

2.2.3 Flight muscle morphological organization

The distinguishing features of insect flight muscle include the large size of the individual fibrils arranged into giant fibres, often of several μm diameter. These fibres are massed to fill the major part of the thorax, and are served by a prolific tracheolar system. Columns of large, closely-packed mitochondria (sarcosomes) lie between the fibrils. In the asynchronous-type flight muscle of the blowfly, *Phormia regina,* for example, these motochondria are up to 4 μm in length and make up some 40% of the total muscle mass (Fig. 2.1). Two interfibrillar membrane systems can be recognized in insect flight muscle. In the synchronous-type muscle, as in most striated muscle of vertebrates, longitudinally oriented sheet-like cisternae of the sarcoplasmic reticulum occur between the fibrils and the mitochondria. Interposed between the mitochondria and these cisternae are transversely oriented tubules of the T-system, which consists, in part, of plasma membrane surrounding the invaginating tracheoles. The transverse tubules come into close contact with the sarcoplasmic reticulum in 'dyad' configuration. Branches of the tracheal system penetrate deep into the fibre allowing gaseous oxygen to reach the vicinity of the mitochondria. Whereas the sarcoplasmic reticulum is markedly diminished in asynchronous flight muscle as compared with synchronous muscles, the T-system is more profusely developed in the former type of muscle. Both types of flight muscle are deeply impregnated with an immense volume of extracellular milieu [5].

Mitochondria from flight muscle are distinctive in that the cristae formed by infolding of the inner membrane are particularly numerous

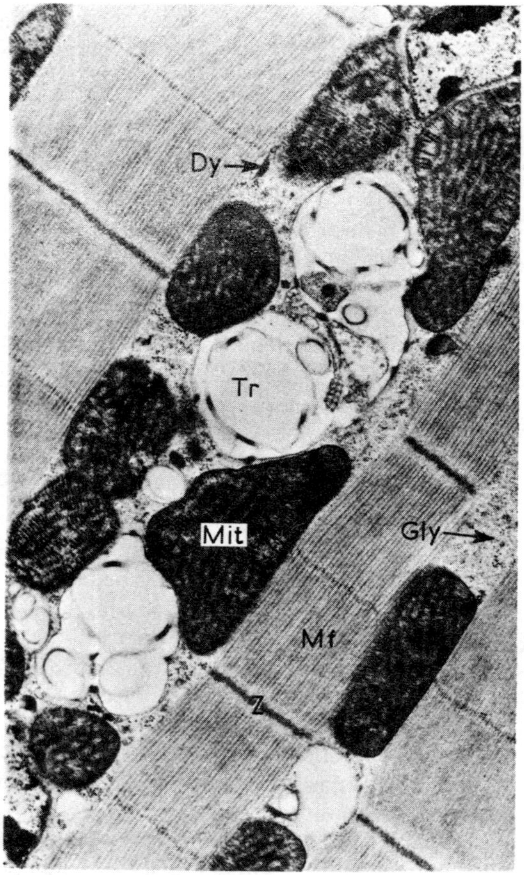

Fig. 2.1 Longitudinal section of an asynchronous-type flight muscle of the blowfly, *Phormia regina* (x 15 000).
Dy–dyad; Gly–glycogen; Mit–mitochondrion; Mf–myofibril; Tr–tracheole; Z–Z-line. (Photograph kindly furnished by Dr. B. Sacktor.)

and are packed very tightly so that they occupy much of the lumen. In rat liver mitochondria, the cristae are sparse. These ultrastructural differences between the mitochondria from blowfly flight muscle and rat liver reflect the high requirement for ATP, produced by the respiratory chain, and the presence of fewer matrix enzyme systems in the insect organelles. Since the outer membrane is generally permeable, the pervasive nature of the flight muscle mitochondrial cristal membrane forms a large effective surface area for rapid exchange of substances between the mitochondrion and cytosol.

Glycerol-3-phosphate dehydrogenase, trehalase and proline dehydrogenase are important enzymes in insect oxidative metabolism, and are all located on the outer surface of the inner mitochondrial membrane, and are, therefore, readily accessible to cytoplasmic metabolites.

It is clear that the flight muscle ultrastructure allows rapid movement of metabolic fuel and oxygen to the oxidative sites within the mitochondria for production of ATP, which can be utilized expeditiously by the adjacent muscle fibres during sustained activity. Whereas the phosphagen of vertebrates is entirely creatine phosphate, only arginine phosphate has been detected in insects, where it occurs in flight muscle in much lower concentrations than the usual phosphagen content of vertebrate muscle. It may be that the efficient aerobic ATP production of flight muscle reduces the necessity for an intermediate energy store [8].

2.3 Carbohydrate metabolism

The two most important carbohydrate energy reserves used by insects are glycogen and the disaccharide, trehalose. Glycogen is usually stored in the flight muscle, fat body and gut. Trehalose is the principal haemolymph sugar in many insect species, but it also occurs in the fat body where it is rapidly synthesized, and in muscle. Although the haemolymph trehalose concentration varies greatly, it is commonly in the range 1–2g/100ml. In contrast, the haemolymph glucose level is negligible. Trehalose clearly occupies a central position amongst the distinctive features of insect carbohydrate metabolism. Closely allied to these specializations are peculiarities in regulatory mechanisms, details of which are largely beyond the scope of this book.

2.3.1 *Interconversion of glucose, glycogen and trehalose*

Although the initial stages of flight may be supported by the limited stores of respiratory fuels in insect flight muscle, maintenance of prolonged flight requires a supply of fuels to the muscles from other tissues, principally the fat body. In blowflies, for example, the fuel during the first two minutes of flight is flight muscle trehalose; thereafter glycogen in the flight muscle is utilized for about 10 minutes, followed by that in the fat body [5]. The glycogen energy reserve is mobilized from the fat body, not as glucose, but as trehalose, which enters the haemolymph for transport to the flight muscle [9]. The interconversions of glucose, glycogen and trehalose are shown in Fig. 2.2.

The control mechanisms of both synthesis and utilization of glycogen and trehalose are closely interrelated [5,9]. In blowfly flight muscle at least, the conversion of phosphorylase *b* to phosphorylase *a* provides a mechanism for controlling the intense rate of glycogenolysis occurring during the initial stages of flight. Since trehalose is also synthesized from UDP-glucose and glucose-6-phosphate in the fat body, control of glycogenolysis in this organ differs somewhat from both the flight muscle and mammalian systems [9]. Glycogenolysis in fat body may ultimately be under hormonal control. A stimulation of the phosphorylase activity, albeit somewhat slow, is produced by a corpora cardiaca (brain-innervated glands) extract, presumed to contain a neurosecretory hormone [10].

[^{14}C]-Glucose, when injected into insects, is rapidly converted into blood trehalose, the principal site of synthesis being the fat body, where the rate of formation and release of this disaccharide is sufficiently rapid to account for most of the trehalose which accumulates in haemolymph and which is utilized during flight [5,11]. The formation of trehalose in the fat body is economical for the insect, since this disaccharide may pass from the cell into the haemolymph by simple diffusion, rather than the energy-consuming process of active transport. After reaching the site of utilization, trehalose diffuses into the cell and is hydrolysed into two molecules of glucose by trehalase, which has been identified in many insect tissues, including flight muscle [11]. Some energy in the form of ATP is required in the synthesis of trehalose.

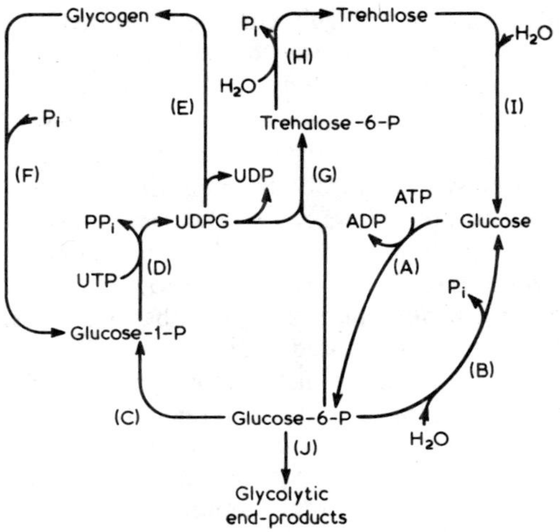

Fig. 2.2 Interconversion of glucose, glycogen and trehalose. A–Hexokinase; B–Glucose-6-phosphatase; C–Phosphoglucomutase; D–UDP-glucose pyrophosphorylase; E–Glycogen synthetase; F–Glycogen phosphorylase; G–Trehalose-6-P synthetase; H–Trehalose-6-P phosphatase; I–Trehalase; J–Glycolysis. Redrawn from [5].

2.3.2 *Catabolism of carbohydrates in flight muscle*

Since insect flight muscle is capable of the most

intense and most highly controlled respiration in nature, it is obviously the system of choice for studies of the underlying mechanisms. Consequently, this discussion will be centred primarily upon investigations of flight muscle, as carbohydrate metabolism in most other insect tissues follows closely the common vertebrate pattern. Emphasis will be placed upon flight muscle metabolism in the blowfly, *Phormia regina*, since this system has been most thoroughly investigated [12].

The metabolic specializations in insect flight muscle are related to the requirement for an exceptionally high rate of ATP production during flight, the consequent high rates of fuel utilization, and the total commitment of flight muscle to aerobic metabolism. This is reflected in the manner in which the NADH formed by the glyceraldehyde 3-phosphate dehydrogenase-catalysed reaction of glycolysis is reoxidized. During glycolysis in Dipteran flight muscle, this NADH is reoxidized principally during the transformation of dihydroxyacetone phosphate into glycerol-3-phosphate, a reaction catalysed by an extremely active cytosolic NAD-linked glycerol-3-phosphate dehydrogenase (Fig. 2.3).

In contrast, the lactate dehydrogenase-catalysed conversion of pyruvate into lactate in the cytoplasm is the only way of reoxidizing NADH in those insect muscles e.g. leg muscles, which are intensely active for short time periods but have long recovery times. Also, organs in which carbohydrate utilization is at a generally low level contain relatively high concentrations of lactate dehydrogenase (LDH). However, muscle which must sustain a high level of activity for long periods, e.g. flight muscle, contains a highly active cytoplasmic NAD-linked glycerol-3-phosphate dehydrogenase (G–3–PD). It has been shown that at the time the locust flight muscle becomes active at the adult moult, the specific activity of G–3–PD increases by a factor of 20 and the G–3–PD/LDH ratio becomes 100 [13]. The glycolytic pathway we have just described,

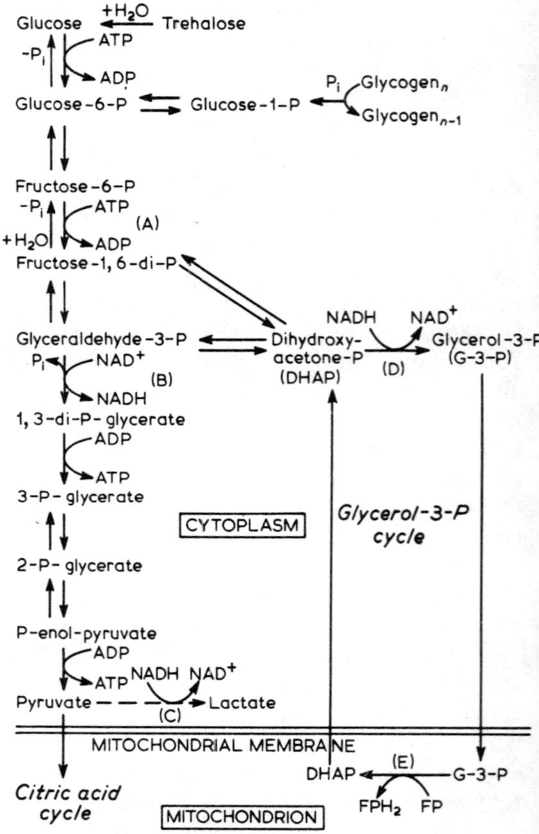

Fig. 2.3 Glycolysis and the glycerol-3-P cycle in flight muscle.
A–Phosphofructokinase; B–Glyceraldehyde-3-P dehydrogenase; C–Lactate dehydrogenase; D–Cytoplasmic glycerol-3-P dehydrogenase; E–Mitochondrial glycerol-3-P dehydrogenase. Modified from [12].

in which pyruvate and glycerol-3-phosphate are end products, is characteristic of the flight muscle of insects, particularly ones with high wingbeat frequencies, such as Diptera and Hymenoptera, which utilize carbohydrate as practically the only fuel during flight. In insect flight muscle, as in vertebrate heart and red muscle in which metabolism is also primarily aerobic, the pyruvate is oxidized via the citric acid cycle.

Many observations led to the original demonstration that mitochondria are impermeable to NADH and NADPH and their oxidized forms [6]. However, the glycerol-3-phosphate formed in the cytoplasm is readily oxidized by the mitochondrial glycerol-3-phosphate dehydrogenase (a flavoprotein) located on the outer surface of the inner mitochondrial membrane thus regenerating dihydroxyacetone phosphate, which then becomes available for further oxidation of extramitochondrial NADH (Fig. 2.3). This cyclic shuttle system allows reducing equivalents from the extramitochondrial pool of NADH to traverse the mitochondrial permeability barrier and then be oxidized by the flavoprotein in the mitochondria, where their subsequent transfer to oxygen via the electron transport chain is coupled to oxidative phosphorylation. Since this is a cyclic process, essentially all the glucose-6-phosphate metabolized during prolonged flight is converted into pyruvate, and is thus made available for complete oxidation in the citric acid cycle. It appears that mitochondrial oxidation of pyruvate and glycerol-3-phosphate occurs during steady state flight in the blowfly, *P. regina*, at the same rates as these compounds are formed. Due to the completely aerobic nature of flight muscle glycolysis, end products do not accumulate wastefully, in contrast to the accumulation of lactate in exercising vertebrate white muscle [12].

In muscles from a number of both vertebrates and invertebrates, the mitochondrial glycerol-3-phosphate dehydrogenase is apparently rate-limiting for the glycerol-3-phosphate cycle. Comparison of the maximum activity of the mitochondrial enzyme *in vitro* with that of phosphofructokinase (a measure of the glycolytic potential of a tissue) indicates that operation of this cycle would probably account for reoxidation of essentially all the NADH produced glycolytically in insect flight muscle, but can account for only a small amount in vertebrate muscle [14].

Sites of regulation of glycolysis and mitochondrial oxidations in blowfly flight muscle *in vivo* have been identified largely by concurrent measurement of the variation in concentrations of numerous metabolites in the muscle upon flight initiation [5]. A decrease in the concentration of hexose monophosphates, accompanied by a rapid accumulation of fructose-1,6-diphosphate (reaching a peak after 15 s of flight) revealed a primary control point at the phosphofructokinase reaction (Fig. 2.4). Like the mammalian enzyme, the phosphofructokinase from insect flight muscle is inhibited by excess ATP, the inhibition being reversed by AMP, cyclic AMP, and P_i. The finding that the concentration of ATP in blowfly flight muscle on commencement of flight decreases, whereas P_i, ADP and, particularly, AMP concentrations increase, is qualitatively in accordance with an enhancement of phosphofructokinase activity. However, the magnitude of the decrease in ATP relative to the increase in AMP is not sufficient to account for the 100-fold increase in the glycolytic rate upon commencement of flight. Thus the activity of phosphofructokinase may be regulated by factors, as yet unknown, but may involve a substrate cycle between fructose-6-phosphate and fructose-1,6-diphosphate [14]. Different responses to certain effectors have been observed not only between phosphofructokinase from locust fat body and flight muscle, but also between the insect and mammalian enzymes [13]. To avoid wastage, it is important for the organism to co-ordinate the citric acid cycle and glycolysis. In mammalian muscle this co-ordination operates through the inhibitory action of citrate on phosphofructokinase, but such an effect could not be demonstrated in the locust flight muscle.

The flight muscle of several species of

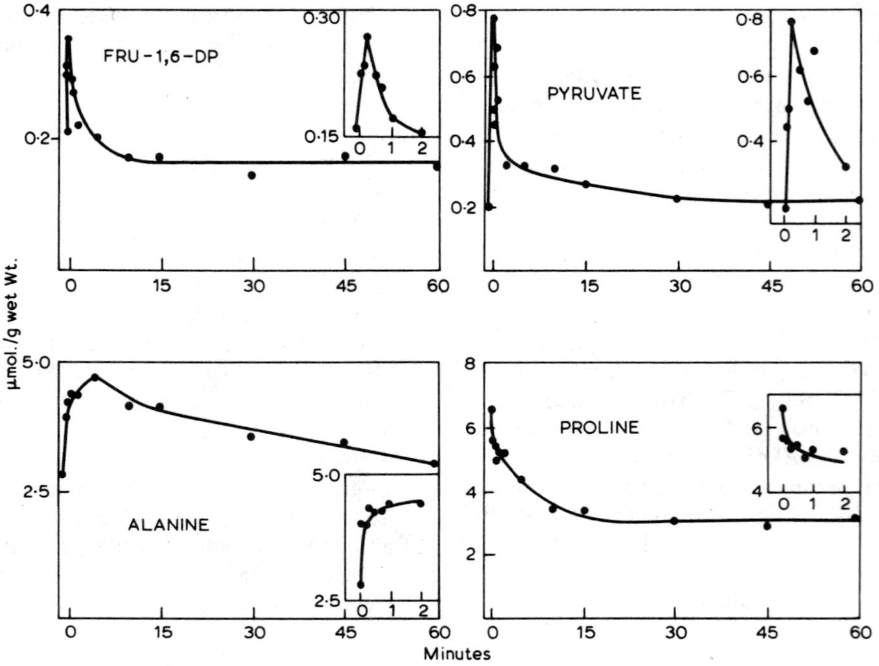

Fig. 2.4 Changes in concentrations of metabolites in thoraces of blowflies during flight. Redrawn from [5].

bumble bee, despite being non-gluconeogenic, is unusual in containing particularly high levels of fructose-1,6-diphosphatase (FDPase), which is involved apparently in heat generation. This enzyme is the only currently known FDPase which is not inhibited by AMP. This allows simultaneous activity of phosphofructokinase and FDPase, leading to a futile carbon cycle at the fructose-6-phosphate $\rightleftharpoons$ fructose-1,6-diphosphate interconversion, with the net result of ATP hydrolysis and production of heat [12].

We can now consider the mitochondrial oxidation of the products of glycolysis. The flight muscle of *Phormia* at rest contains an exceptionally high concentration of glycerol-3-phosphate, which remains essentially constant during flight. This indicates that the additional glycerol-3-phosphate formed during oxidation of the extra NADH produced by increased glycolytic flux is immediately oxidized. Together with other evidence, this suggests that mitochondrial glycerol-3-phosphate oxidation is activated on transition of the muscle from rest to activity. These studies *in vivo* also demonstrated a dramatic rise in the concentration of pyruvate and a concomitant accumulation of alanine and acetyl carnitine on transition from rest to flight (Fig. 2.4); the pyruvate concentration returned to a low steady-state level by the second minute of flight. This suggests that initially pyruvate is not oxidized in the citric acid cycle as quickly as it is formed by glycolysis, and that constraint on pyruvate oxidation is released shortly after the onset of flight. There is also a coincident and stoichiometric relationship between the rise in alanine and a decrease in proline, which occurs in the resting muscle in extraordinarily high concentrations (Fig. 2.4). This suggests that the amino group in alanine is mainly derived from proline [5].

To gain some insight into the significance of proline mobilization, we must first consider the permeability properties of intact blowfly flight muscle mitochondria [6]. It is apparent that blowfly flight muscle mitochondria oxidize only exogenous glycerol-3-phosphate, pyruvate, acetyl carnitine and, to a lesser extent, proline at appreciable rates *in vitro.* None of the citric acid cycle intermediates and neither glutamate nor aspartate is an effective respiratory substrate when added to the isolated mitochondria. This is due to the unusual phenomenon that insect flight muscle mitochondria are not readily permeable to these compounds. With one or two exceptions, it has not been possible to demonstrate the presence of specific dicarboxylate and tricarboxylate anion translocases (permeases) for exchange transport of citric acid cycle intermediates, analogous to the permeases found in mammalian mitochondria. In fact, the interesting suggestion has been made that this may be an adaptation to prevent loss of the citric acid cycle intermediates from the mitochondria, rather than merely to prevent their entry. This would be advantageous in flight muscle, where the citric acid cycle intermediates are required almost exclusively for respiratory purposes, and not for their involvement in extramitochondrial biosynthetic processes, which occur in tissues such as mammalian liver [6].

We are now in a position to consider the function of proline in mitochondrial oxidations [12]. Isolated insect flight muscle mitochondria quickly lose the capacity to oxidize pyruvate. In blowfly mitochondria, this loss can be reversed by proline, but not by citric acid cycle intermediates nor glutamate. This is in agreement with the permeability properties and the ability of such mitochondria to oxidize exogenous proline but not the other metabolites tested. These findings led to the suggestion that the physiological significance of proline oxidation in blowfly flight muscle *in vivo*, particularly at the initiation of flight, is to provide citric acid cycle intermediates required for complete oxidation of pyruvate as quickly as it is formed. Proline oxidation by flight muscle preparations apparently occurs in all the major orders of insects, the pathway being the same as that in mammalian tissues with the formation of glutamate via Δ′-pyrroline-5-carboxylate (Fig. 2.5). Transamination of this glutamate with pyruvate yields alanine, which is further metabolized via the citric acid cycle to oxaloacetate. This can then condense with acetyl CoA, produced from pyruvate, to yield citrate, thereby effecting the complete oxidation of pyruvate via the citric acid cycle. The studies on metabolite levels in

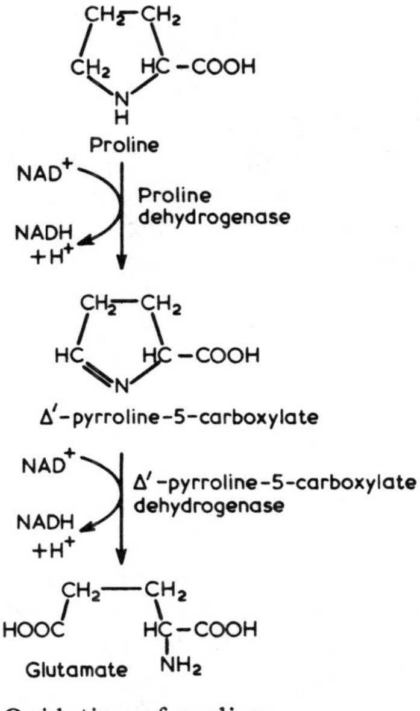

Fig. 2.5 Oxidation of proline.

blowfly during rest and flight, referred to earlier, are also in accord with this concept. Presumably only when adequate oxaloacetate has been produced from proline oxidation can pyruvate be oxidized in the citric acid cycle at the same rate as it is produced by glycolysis. In the meantime, unoxidized pyruvate either accumulates, or transaminates with glutamate to give alanine, or yields acetyl CoA, which forms acetyl carnitine or acetate. The apparently unique role of proline in furnishing precursors of oxaloacetate has yet to be explained satisfactorily.

Whereas in the blowfly the suggested physiological importance of proline oxidation in flight muscle has been rationalized in terms of provision of citric acid cycle intermediates during flight initiation, this explanation cannot be applied universally, as oxidation of proline is apparently of more importance in flight muscle metabolism in the tsetse fly, *Glossina morsitans.* Bursell [15] showed that proline concentration falls sharply during flight in this insect, with a concomitant and practically stoichiometric rise in alanine level. Labelling studies were consistent with the operation of the metabolic cycle shown in Fig. 2.6. Conversion of one mole of proline into one mole of alanine by this scheme can produce 14 moles of ATP, which compares favourably with the 15 moles produced by complete oxidation of one mole of pyruvate via the normal citric acid cycle. The finding that glycogen is barely detectable in the tsetse fly, with trehalose and glucose present in only minimal amounts, suggests that proline oxidation and ancillary pathways are of central importance in this insect. Also in accordance with this, the enzymes of carbohydrate catabolism, including phosphorylase, trehalase, phosphoglucomutase, hexosephosphate isomerase, phosphofructokinase, and aldolase have low activities in tsetse fly flight muscle, compared to the blowfly. This argues for a minimal role of glycolysis in the tsetse fly. Succinic dehydrogenase, which is required in the proposed tsetse fly scheme (Fig. 2.6), has similar activity to that in the blowfly. The important role of proline in tsetse fly metabolism may be an adaptation to the insect's exclusive feeding on vertebrate blood and consequent ingestion of insignificant amounts of carbohydrate. However, haemoglobin contains proline which may be present at a level of 2% in the haemolymph of resting insects. The strikingly high activities of proline dehydrogenase in the flight muscles of the Colorado potato beetle, *Leptinotarsa decemlineata,* and the cockchafer, *Melontha melontha,* may suggest that emphasis on proline oxidation need not necessarily be of limited occurrence [12].

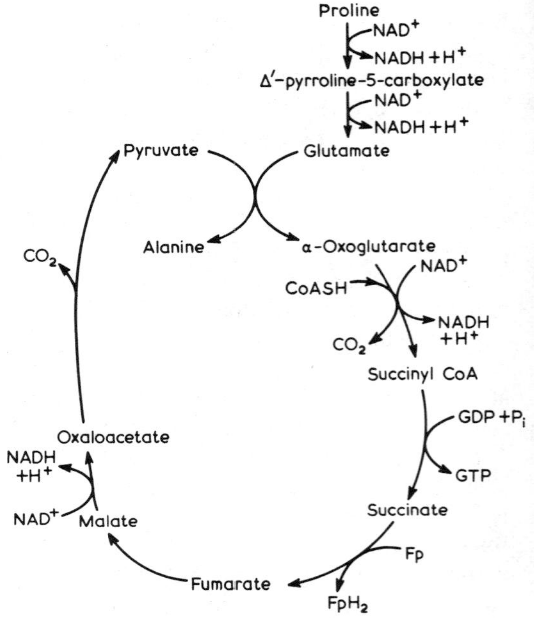

Net reaction: Proline $\longrightarrow$ Alanine $+ 2CO_2 + 3NADH + 3H^+ + 2$ Flavoprotein. H_2 + GTP

i.e. Proline $\longrightarrow$ Alanine $+ 2CO_2 + 5H_2O + 14$ ATP

Fig. 2.6 Proline oxidation during flight in the tsetse fly, *Glossina morsitans.*

2.3.3 The respiratory chain

Studies on insect flight muscle mitochondria have greatly contributed to advance our general understanding of electron transport and oxidative phosphorylation. The advantages of the insect flight muscle mitochondria as a suitable experimental system are largely due to the generally high level of oxidative metabolism, the high degree of respiratory control, and the relatively high content of cytochromes. It is significant historically that the general concept of a respiratory chain was first proposed by Keilin in 1925 from observations on insect flight muscle.

When the difference spectrum of reduced and oxidized mitochondria from housefly flight muscle is examined at $-190°C$, which results in significant sharpening and intensification of the cytochrome absorption bands, at least two major differences in cytochrome composition from the pattern in mammalian mitochondria become apparent [6].

Housefly mitochondria reduced either enzymically or non-enzymically had no absorption band at 554 nm, a characteristic peak of cytochrome *c*, which is a major component of the respiratory chain of mammalian mitochondria. However, in preparations from which the bulk of the cytochrome *c* had been removed by washing with saline solution in order to eliminate any masking effect, a component with an α band at 551 nm became visible. This cytochrome in housefly mitochondria (and probably also in others) which absorbs at 551 nm at $-190°C$, is apparently analogous in function to mammalian cytochrome *c*.

The second major distinctive feature of the insect system was revealed by reduction of the housefly flight muscle mitochondrial pigments either non-enzymically with dithionite or enzymically in the presence of antimycin A, when a peak at 555 nm was observed on examination at $-190°C$. A suggested sequence of the various respiratory chain components based on kinetic studies and on the effects of specific inhibitors is shown in Fig. 2.7 [16]. The component designated cytochrome 555 appears on addition of antimycin A or dithionite, but is not reduced during substrate-induced anoxia. Although this cytochrome is located between substrate and the antimycin A-sensitive step, it is shown separate from the electron-transport chain, since its function is uncertain.

2.3.4 Control of flight muscle mitochondrial oxidations and respiration

The respiratory control ratio is defined as the ratio of the rate of oxygen uptake in the presence of added ADP to the rate of respiration after the added ADP has been completely used. Fly flight muscle mitochondria are particularly suitable for studies on control of oxidations because they exhibit high respiratory control ratios and thus show large changes in flux through the citric acid cycle, and also because of the simplicity imposed by the lack of mitochondrial permeability to cycle intermediates. In fact the citric acid cycle in blow-

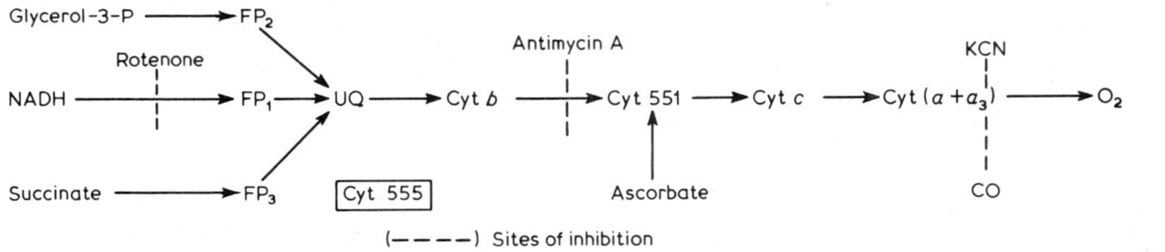

Fig. 2.7 Possible arrangement of the components of the respitory chain. Redrawn from [16].

fly flight muscle essentially serves a purely catabolic function.

Chance and Williams proposed a classical scheme for the control of mitochondrial oxidative metabolism, in which the respiratory activity is essentially determined by the relative concentrations of ATP, ADP, and phosphate [17]. When the concentrations of ADP and phosphate are high but the ATP concentration low, the respiratory rate is maximal. Conversely, when the concentrations of these effectors is reversed, a low rate of respiration is observed. In the absence of ADP, the transfer of electrons along the respiratory chain is inhibited, and the NAD becomes reduced, with a resulting inhibition of dehydrogenase activity. This scheme has been found to be applicable to mitochondria from diverse sources.

Experiments with blowfly flight muscle mitochondria *in vitro* showed that ADP significantly increased the respiratory rate with either glycerol-3-phosphate or pyruvate as substrate. However, there is extensive evidence that respiratory control in flight muscle may be distinctive in that control at the substrate (dehydrogenase) level is apparently more important than release of inhibition of the respiratory chain by ADP in the classical manner proposed by Chance and Williams [12,16].

The role of various metabolites acting at the dehydrogenase level in the control of respiration of insect flight muscle mitochondria will now be considered. Evidence has already been noted suggesting that mitochondrial oxidation of glycerol-3-phosphate is activated on transition of the muscle from a resting to an active state. Ca^{2+} lowers the apparent K_m (i.e. increases affinity) of the mitochondrial glycerol-3-phosphate dehydrogenase for the substrate. It is envisaged that regulation of this enzyme is achieved by release of Ca^{2+} from the sarcoplasmic reticulum simultaneously with nervous stimulation of the muscle at the initiation of flight, thus reversing the inhibited state. In fact, there is evidence that in the resting state, the dehydrogenase is probably the limiting factor for the oxidation of glycerol-3-phosphate. Some supplementary increase in the rate of glycerol-3-phosphate oxidation by ADP interaction with the respiratory chain during flight initiation is not precluded.

The fact that the particularly high concentration of proline in flight muscle of the blowfly, *P. regina*, drops rapidly on initiation of flight led to the suggestion that mitochondrial oxidation of proline is facilitated by the rest-to-flight transition. This oxidation is stimulated by ADP in the presence of uncoupling agents, indicating that the site of action of ADP is on proline dehydrogenase, rather than on the respiratory chain. ADP is an allosteric effector of the dehydrogenase and acts by lowering the apparent K_m for proline to a value approximating to the concentration of proline found in the muscle at rest. The finding that pyruvate also increases the affinity of proline dehydrogenase for its substrate, coupled with the fact that there is a dramatic increase in the flight muscle pyruvate concentration upon initiation of flight, suggests an additional way of stimulating proline oxidation at the level of the dehydrogenase. Since proline dehydrogenase is inhibited by ATP, the activity of the enzyme is dependent on the relative proportions of the adenine nucleotides as well as on the absolute level of ADP. Since the enzyme is particularly sensitive to small increases in ADP even in the presence of a large amount of ATP, the small changes in ADP and ATP levels in blowfly flight muscle on initiation of flight will lead to a considerable enhancement in the rate of proline oxidation.

We have already considered a possible mechanism for control of the mitochondrial oxidation of pyruvate and of the citric acid cycle at the beginning of flight arising from a deficiency in cycle intermediates. This deficiency is relieved when precursors of oxaloacetate, derived from proline oxidation, become available. Other mechanisms of control of pyruvate oxidation will now be considered. Several lines of evidence support the concept that a dehydrogenase, rather than the respiratory chain, limits the rate of oxidation of pyruvate by blowfly mitochondria in the absence of ADP. It is interesting that the converse apparently holds for mitochondria from all other classes of animals yet investigated. In the insect system, there is good evidence that NAD-linked isocitrate dehydrogenase is the principal rate-limiting enzyme [18]. However, supplementary control loci of pyruvate oxidation may occur at pyruvate dehydrogenase and α-oxoglutarate dehydrogenase. As in the case of glycerol-3-phosphate oxidation, concurrent activation of the respiratory chain by effectors (ADP, P_i) is not excluded. Flight muscle NAD-isocitrate dehydrogenase is controlled by a number of effectors. Amongst these, the phosphorylation state of the intramitochondrial adenine nucleotides is important; this also controls the activity of α-oxoglutarate dehydrogenase.

In some mammalian tissues, control of pyruvate oxidation occurs at the level of the pyruvate dehydrogenase complex. Pyruvate dehydrogenase kinase, in the presence of ATP and Mg^{2+} converts the active enzyme into its inactive phosphorylated form; reactivation of the enzyme occurs by the action of a specific phosphatase. Both enzymes occur in blowfly flight muscle. The kinase is inhibited by pyruvate, so that when the pyruvate concentration rises, as during flight initiation, its oxidation is facilitated; it is also significant that the phosphatase is activated by Ca^{2+} [12].

The changes in the concentrations of effectors upon flight initiation, in general, conform with the postulated mechanisms for regulation of the mitochondrial oxidations as well as of glycolysis and glycogenolysis.

2.3.5 Metabolic changes during development

In view of the profound changes occurring during metamorphosis of holometabolous insects, it is not surprising that there is a wealth of detailed analytical data on the changes in various biochemical parameters during insect development [19–21]. During development of the flight muscle, there is a tremendous increase in the mitochondrial volume, which continues into the adult stage [22]. Developing flight muscle is an excellent system for study of mitochondrial biogenesis, particularly in view of the large changes involved. It appears that mitochondrial growth is not autonomous, and is dependent upon contribution from extramitochondrial ribosomes [23].

Isoenzymes (distinct and separable proteins possessing the same catalytic activity) have great potential in investigations of development and morphogenesis. Isoenzyme patterns not only generally change during the development of an organism, but also show a high degree of specificity at the tissue (or cell-type) level, and at the subcellular level. Glycerol-3-phosphate dehydrogenase activity in *Drosophila melanogaster* provides a good example of these three types of isoenzyme pattern. The vast amount of genetic information which has been amassed for *Drosophila* makes this an animal of choice for developmental studies, although the small size of the organism may be a serious disadvantage for certain types of biochemical investigations. In *Drosophila*, isoenzymes have been used, for example, to determine at what stage in development particular genes are turned on, and also to establish cell lineages during development of imaginal discs (groups of cells destined to form adult structures) [24].

2.3.6 Metabolism during diapause

Diapause is a state of developmental arrest with the ecological significance of tiding the animal over the winter or other unfavourable season. This phenomenon, which is characterized by exceedingly depressed rates of metabolism, may occur at the egg, larval, pupal or adult stages depending upon the insect species. Diapause is under endocrine control, the nature of which differs with the stage and species.

During the overwintering stage of many insects, glycogen is converted into polyhydric alcohols, which subsequently can be reconverted into glycogen. For example, the diapause embryo of the commercial silkworm, *Bombyx mori*, produces sorbitol and glycerol, and the *Hyalophora cecropia* (cecropia silkmoth) pupa produces glycerol. Whereas glycerol production in these species coincides with the onset of diapause, glycerol production can also occur in non-diapausing insects, as shown by its formation in adult carpenter ants (*Camponotus pennsylvanicus*) and Alaskan carabid beetles (*Pterostichus brevicomis*) on exposure to cold. The level of glycerol can be particularly high; a value about 25% of the body weight has been reported for diapause prepupae of *Bracon cephi.* The accumulation of glycerol and sorbitol is apparently of some ecological significance in protection against cold. They probably act as an 'antifreeze' by lowering the supercooling point. Not all cold-resistant insects contain high concentrations of glycerol, whereas many cannot survive even with a large amount [25].

Early observations particularly on the silkmoth, *Hyalophora cecropia,* indicated that the respiratory rate of diapausing pupae is not only dramatically reduced, but it is also highly resistant to the cytochrome oxidase inhibitors, carbon monoxide and cyanide [21,26]. However, tissues in diapause do contain low levels of cytochrome oxidase, although their cytochrome content is much reduced. The tissues apparently contain more cytochrome oxidase than required to maintain the low rate of electron transport during diapause, so that the oxidase can be inhibited to a considerable extent without affecting net pupal respiration. However, inhibition can be demonstrated when the demand upon the cytochrome oxidase is increased by either enhancing the rate of respiration (e.g. with 2,4-dinitrophenol) or by lowering the oxygen pressure. The fact that dinitrophenol strongly stimulates oxygen uptake by diapause pupae indicates that a limitation of cytochrome components is not responsible for the low rate of diapause respiration. Respiratory control in this case also probably occurs via the usual coupled phosphorylation system, and the diminished respiratory rate apparently results from a reduced energy demand during diapause.

Early investigations on the embryonic diapause of silkworm (*Bombyx mori*) revealed the occurrence in the eggs of enzymes catalysing the following reactions:

Glucose + NADPH + $H^+ \rightleftarrows$ Sorbitol + $NADP^+$
Glucose-6-P or F-6-P + NADPH + $H^+ \rightleftarrows$ Sorbitol-6-P + $NADP^+$
Glyceraldehyde or dihydroxyacetone + NADPH + $H^+ \rightleftarrows$ Glycerol + $NADP^+$
Dihydroxyacetone-P + NADH + $H^+ \rightleftarrows$ Glycerol-3-P + NAD^+.

Details of the pathways for formation of both sorbitol and glycerol and their control are still lacking, although a clearer picture for production of the latter is gradually emerging.

In the diapausing pupae of the silkmoth (*H. cecropia*), although some glycerol production occurs at 25°C, the rate of production is greatly enhanced at lower temperature (6°C). The glycerol is mainly formed at the expense of fat body glycogen, and enzymes catalysing a plausible pathway occur in insects: glycogen phosphorylase, the glycolytic pathway to dihydroxyacetone phosphate, glycerol-3-phosphate dehydrogenase and glycerol-3-phosphatase. Glycogen phosphorylase in these pupae is activated by cold; this reversible temperature effect probably reflects interconversion of enzyme forms by phosphorylase kinase and phosphatase [27]. Thus, glycogenolysis in diapausing cecropia silkmoth pupae is regulated in part by the environmental temperature. However, the specific formation of glycerol is probably controlled at the glycerol-3-phosphatase step.

2.4 Lipid metabolism

Dietary lipids can be deposited in the fat body, which is also capable of lipid synthesis from non-lipid precursors. The fat body triglycerides can be mobilized in various forms and transported largely as constituents of haemolymph lipoproteins to the muscles where they function as fuel. Although it had been suggested that this lipid is released from the fat body as diglyceride [28], it now appears that it can be mobilized in the form of triglyceride, diglyceride and fatty acids, depending upon the insect species. Lipid is also oxidized in the fat body to furnish energy to support functions such as gluconeogenesis [9]. Lipid mobilization from the fat body is controlled, at least in some species of insects, by an adipokinetic hormone of the corpora cardiaca (see p. 31). Juvenile hormone also affects lipid metabolism in an, as yet, unclear manner.

On the whole, metabolism of lipids in insects follows routes established for bacteria and vertebrates [29]. Insects, in common with mammals, cannot synthesize C_{18} polyunsaturated fatty acids, thus explaining their nutritional requirement for linoleate or linolenate. In general, insects can elongate and desaturate existing saturated fatty acids but cannot elongate (or saturate) existing unsaturated fatty acids. The major difference in fatty acid biosynthetic capacity between insects and vertebrates is that unlike vertebrates, insects are unable to elongate unsaturated fatty acids or introduce a second or third double bond into any fatty acid. Fatty acids once synthesized are readily incorporated into glycerides, phospholipids and sterol esters. Most insects cannot synthesize choline, and therefore, require a dietary supply for synthesis of phospholipids and also acetylcholine.

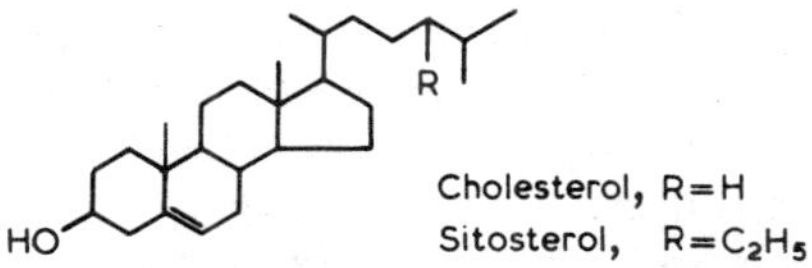

Fig. 2.8 Chemical structures.

Insects, in common with many other invertebrates, cannot synthesize sterols *de novo*; there is extensive deletion of the pathway after farnesyl pyrophosphate. However, they must have adequate dietary sterol for normal growth and development, and practically all insects have an obligatory requirement for a minimal amount of cholesterol. This is derived directly from the diet in carnivorous insects, but is obtained by dealkylation of major plant sterols, such as sitosterol (Fig. 2.8), in phytophagous insects [30]. Although some of the cholesterol is required as a precursor of the insect moulting hormones, most of it fulfils a specific structural role in membranes. It is significant that phytophagous insects, in common with many other arthropods, have evolved this metabolic capability for formation of their obligatory cholesterol by dealkylation of plant sterols; carnivorous insects lack this metabolic capacity. This represents an interesting evolutionary adaptation of phytophagous insects to their ecological niche.

Carotenoids, important in pigmentation in many insects, are taken up from the plant diets by phytophagous insects and in many cases metabolized by oxidation; carotenoproteins also frequently occur [31]. β-Carotene is also involved in formation of visual pigments in insects [2].

It appears that the capacity of flight muscles for high power output and the choice of specific fuel for flight (carbohydrate v lipid) are based upon quantitative changes in common enzyme systems, in conjunction with adaptations in the specific permeability properties of the sarcosomes. Thus, most citric acid cycle intermediates essentially cannot pass out of fly muscle sarcosomes and are available for a high level of carbohydrate oxidation. The activities of the enzymes, lipase and carnitine-acyltransferase (involved in transport of acyl residues across the mitochondrial membrane) show overall correlation with the ability of a particular flight muscle to utilize lipid as a predominant fuel for flight.

2.5 Nitrogen metabolism

2.5.1 Amino acid and protein metabolism

The free amino acid concentration in insect haemolymph is relatively high, often being about 50 times that of vertebrate blood. However, the exact significance of this is uncertain. The fat body is an active site for intermediary metabolism of amino acids in insects, which require the same 'essential' amino acids as mammals, with the addition of arginine. Although no marked differences in amino acid metabolism have been reported in insects, some additional pathways of amino acid transformation into heterocyclic products occur [32]. For example, tryptophan serves as a precursor of the ommochrome group of pigments [33]. The ommochrome pigments produce the deep colour of insect eyes and contribute to the brilliant body colouration in many species. Elegant experiments demonstrated that mutants of *Drosophila* and *Ephestia* with abnormal eye pigments are genetically blocked at different points along the pathway to the ommochromes. This work was of considerable historical significance, since it contributed substantially to the development of the concept that genes control the synthesis of enzymes, which in turn are responsible for specific metabolic reactions.

Another group of pigments, the pteridines, which were first isolated from the wings of butterflies, are formed from purines. Besides being important as body pigments in Hymenoptera and Lepidoptera, and functioning as accessory pigments in insect eyes, they occur throughout the plant and animal kingdoms, and are of general importance in cellular metabolism, e.g. as cofactors of enzymic reactions [32].

The physiological role of the fat body is very much dependent upon the developmental stage and the actual time within instars. For example, in *Calliphora erythrocephala* larvae, the fat body in early third instar larvae synthesizes and releases protein into haemolymph, whereas in late larvae it sequesters protein back from the haemolymph [34]. Whereas protein is stored in the larval fat body to be used later in the assembly of adult tissues at the time of metamorphosis, protein storage in the adult fat body is associated with the reproductive functions of the adult stage. The fat body in female insects synthesizes yolk proteins (vitellogenins), which are afterwards translocated to the developing oöcytes. The fat body at all stages appears to synthesize much haemolymph protein including enzymes.

2.5.2 Nitrogen excretion

The Malpighian tubule-rectum complex is the main excretory system in insects. Although historically uric acid has been regarded as the most important single nitrogenous waste material in terrestrial insects, other closely related compounds which are all involved in the pathways of biosynthesis or breakdown of uric acid are also excreted by one or another insect species, and may indeed predominate [35,36]. Some insects, in addition to excreting uric acid have evolved an alternative mechanism of internal deposition of this waste product in the fat body or even the cuticle. There is increasing evidence that stored urates are in equilibrium with metabolic pools, and thus, serve as carbon and nitrogen reserves.

Urea is a minor nitrogenous excretory product in insects. Although the complete ornithine cycle has not been documented for any insect species, many of the reactions associated with the vertebrate cycle occur in insect tissues; enzymes are found for the transformation of arginine into proline. The significance of this awaits complete elucidation.

It has been traditional to assume that a correlation exists between the nitrogenous excretory products of animals and their habitats; aquatic organisms usually excrete toxic ammonia, whereas their terrestrial counterparts may excrete other substances requiring less water elimination. However, it now appears that ammonia can be a predominant nitrogenous excretory product even in terrestrial arthropods. For example, in the American cockroach, ammonia can account for as much as 90% of the total nitrogen voided under certain dietary regimes.

2.6 Haemolymph

There is a striking evolutionary tendency for the contribution of inorganic cations and anions to the osmotic pressure of insect haemolymph to be replaced increasingly by a contribution from small organic molecules. For example, in the most primitive insect examined, the apterygote *Petrobius maritimus*, Na^+ and Cl^- still account for nearly the entire total osmotic pressure of the haemolymph, as they do in other arthropods. Lepidoptera, Hymenoptera, and many Coleoptera exhibit the most specialized type of haemolymph, in which the importance of inorganic cations and anions is considerably reduced, the main role as osmolar effectors being played by small organic molecules, such as free amino acids. These groups are also apparently highly specialized in that Na^+ concentrations are very low, whereas those of Mg^{2+} and K^+ are particularly high. The true adaptive significance of a specialized cationic pattern in the most highly evolved insects is not entirely clear [37].

Exceptionally large amounts of organic acids, particularly intermediates of the citric acid cycle can occur in insect haemolymph, where they may function in cationic balance. Other interesting peculiarities of insect haemolymph include a very pronounced amino acidaemia, the occurrence of a high concentration of organic phosphates, and a surprisingly large number of haemolymph enzymes. Although it is possible to attribute physiological roles to some of these enzymes, the precise significance of most of them requires further elucidation.

Whereas in higher vertebrates specific antibodies consisting of γ-globulin proteins are produced by lymphoid tissue, such substances are, according to available evidence, not produced in the lower vertebrates and invertebrates [38,39]. However, the cellular and humoral defence mechanisms of insects are homeostatic adaptations, which effectively recognize and combat not only foreign micro-organisms and parasitoids, but also abnormal or degenerating host tissues. Haemocytes (blood cells) are involved in phagocytosis of small particles and encapsulation of larger organisms [40]. Some insects produce humoral substances (antibacterial, bactericidal, lytic etc.) against various microbial parasites, that provide some degree of immunity. However, these substances are of unknown origin and constitution, and unlike vertebrate antibodies, they lack specificity.
A central question in considering these insect defence mechanisms is the need for elucidation of the recognition factors by which an organism can distinguish 'self' from 'foreign'. Although the molecular basis for destruction of invading pathogens in invertebrates apparently differs appreciably from that in vertebrates, it is probably no less diverse.

2.7 The integument

The insect integument consists of a single layer of epidermal cells together with the cuticle which lines the surface of the animal.

The insect integument is of much interest from a developmental point of view, since the activities of the single layer of epidermal cells change during the development of the insect.

This changing activity is visibly recorded by the synthesis of the distinctly different chemical components of the cuticular layers in definite sequence. The hormonal controls responsible for such changes in activities are of particular interest (Chapter 3). The insect cuticle also has merits as an experimental system for studies on polarity and patterns, since the orientation of external components reflects the polarity of the single underlying layer of epidermal cells. This topic has been considered elsewhere in this series [41], and reviews are also available [42,43].

2.7.1 Structure of the integument

The cuticle, in general, consists of distinct layers. Outside the flexible endocuticle is the exocuticle, which becomes hardened and darkened in certain regions by incorporation of sclerotized proteins to form tough, rigid sclerites. The exocuticle is covered in turn by a thin, complex epicuticle, which is responsible for the extensive impermeability to water of the cuticle in terrestrial insects.

Cuticular protein, which occurs in all the layers, is heterogeneous with the greater majority being covalently bound to chitin, thus forming a glycoprotein complex. Chitin may account for more than half the dry weight of the cuticle, and occurs in both the endocuticle and exocuticle. It is a high molecular weight polysaccharide consisting predominantly or (possibly) entirely, of unbranched chains of β-(1,4)-linked N-acetyl-D-glucosamine residues. Enzymes in the moulting fluid are capable of degrading chitin to N-acetylglucosamine units. At least in outline, the chitin biosynthetic route in insects appears to be the same as that elucidated in the mould, *Neurospora crassa*, and occurs via UDP-N-acetyl-glucosamine. Moulting hormone (ecdysone) is, apparently, involved in controlling the changing sequence of chitin-synthesizing and catabolizing systems. For a detailed account of cuticular components see [44].

2.7.2 The moulting process

During moulting, the epidermis retracts from the old cuticle (apolysis), undergoes mitosis, and produces some new cuticle prior to secreting into the space between the old and new cuticle (ecdysial space) hydrolytic enzymes for digestion of most of the chitin and protein of the old cuticle for resorption and re-utilization. The undigested old cuticle (exuvium) splits and is shed (ecdysis), so that the insect expands in its temporarily elastic covering. In many cases, the new cuticle is hardened within a few hours by cross-linking of the protein in the exocuticle by quinone (sclerotization). The epidermis continues to deposit protein and chitin in the endocuticle during much of the intermoult period.

2.7.3 Sclerotization and puparium formation

Sclerotization (or tanning) is the process whereby the soft, pale cuticle of the newly ecdysed insect becomes hardened and usually darkened over a period of a few hours. This process also occurs during puparium formation which is unique to the higher Diptera. Tanning occurs by interaction of *ortho*quinones, derived mainly from tyrosine, with –NH and –SH groups of cuticular proteins, the resulting hard, high molecular weight skeletal protein being called 'sclerotin'. This cross-linking of peptide chains produces a network of protein and chitin [42, 44].

In the higher Diptera, the fully-grown larva contracts into a barrel-shaped form and the epidermis brings about sclerotization of the larval cuticle to form a hard, dark shell or puparium for protection of the future delicate pupa. A cuticle of larval type (pre-pupal cuticle) is formed within the puparium before the pupa is formed, but the puparium is not shed until imaginal ecdysis (adult emergence) occurs.

Control of sclerotization during puparium formation by ecdysone has been utilized as a convenient system for detailed investigations by Karlson, Sekeris, and associates on the biochemical mode of action of the hormone (Chapter 3). In early third instar larvae of the blowfly, *Calliphora erythrocephala*, tyrosine metabolism occurs mainly by transamination to p-hydroxyphenylpyruvic acid, which is further degraded via the lactic and propionic acid derivatives (Fig. 2.9). However, late in the instar, during the onset of puparium formation, tyrosine metabolism shifts so that the main metabolic pathway becomes production of N-acetyldopamine via DOPA and dopamine, for utilization in the sclerotization process. This anabolic shift correlates temporally with a high titre of ecdysone, which apparently induces the *de novo* synthesis of the controlling enzyme in this pathway, DOPA decarboxylase. A phenoloxidase enzyme possessing both mono- and diphenoloxidase activities is involved in this pathway at two points. This enzyme occurs in a proenzyme form, and the activator protein is claimed to be under ecdysone control. Activation of the prophenoloxidase is also under metabolic control, being inhibited by the phenylcarboxylic acids produced by tyrosine catabolism. After the ecdysone-mediated switch in tyrosine metabolism, the concentrations of these acids decreases, and the activity of the prophenoloxidase activating system is increased [45].

Puparium formation is not only controlled by ecdysone; the occurence of yet another two neurohormones which are obligatory for the process has been established by Fraenkel and co-workers in the fleshfly, *Sarcophaga bullata*, both being distinct from moulting hormones, ecdysiotropic (brain) hormone and bursicon (see below) [46]. A tentative hypothesis for the complex interactions in the control of puparium formation in fleshflies has been suggested (Fig. 2.10). According to this hypothesis, the longitudinal contraction to form a barrel shape, the cuticular shrinkage to smoothen the puparium, and the retraction of the anterior segments are under direct neural control of the central nervous system (CNS). Retraction of the anterior segments is also controlled indirectly through a neurosecretion from the CNS, which

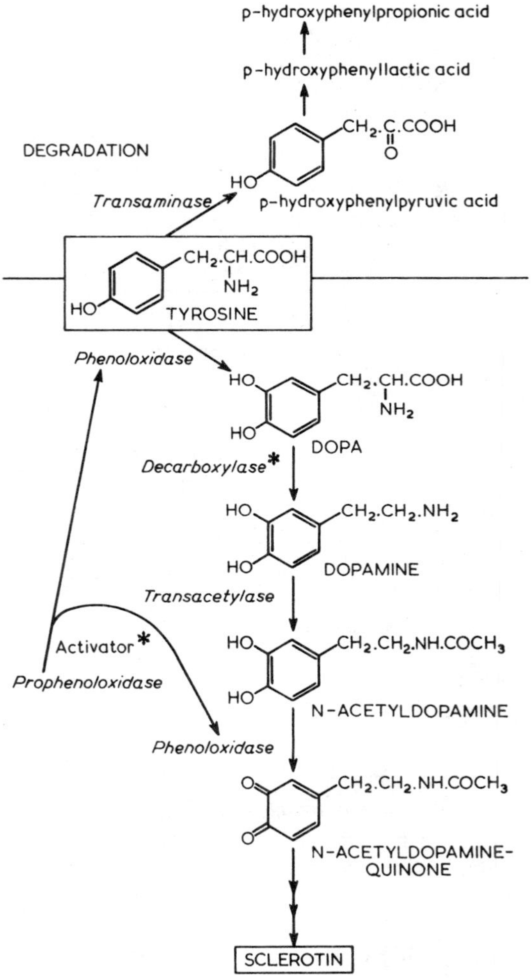

Fig. 2.9 Tyrosine metabolism in *Calliphora erythrocephala*.
*Enzymes under ecdysone control.

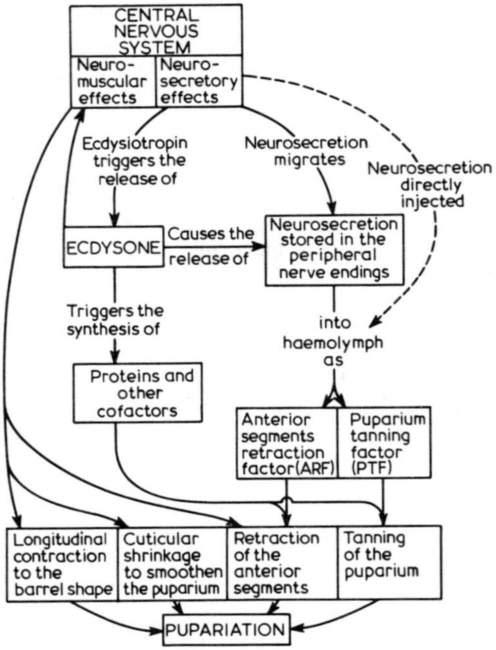

Fig. 2.10 Hypothetical scheme for the control of puparium formation in *Sarcophaga bullata*. Redrawn from [46].

in addition regulates tanning. This neurosecretory material synthesized in the brain is composed of the anterior retraction factor (ARF) which controls the retraction of the anterior segments and the puparium tanning factor (PTF) which regulates tanning. These proteinaceous neurosecretions are presumably stored in the peripheral nerve endings in all segments of the larva, and their release into the haemolymph is triggered by moulting hormone, whose synthesis and/or release is itself controlled by another neurosecretion, ecdysiotropin. Moulting hormone also induces the synthesis of various materials such as nucleic acids, enzyme proteins, tanning substrates and other co-factors which permit the released ARF and PTF to bring about retraction of the anterior segments and tanning. These two neurosecretory factors apparently cannot replace ecdysone in effecting pupariation, and they require ecdysone-dependent enzyme proteins or other co-factors. Thus, during puparium formation, ecdysone, ARF and PTF apparently act in concert to co-ordinate puparial contraction and tanning.

Cuticular tanning (sclerotization) also occurs in newly moulted nymphs and in newly emerged adults of many species of insects. Although a number of diphenols have been implicated in the tanning process, N-acetyldopamine quinone is the major tanning agent in insects of all stages. It is significant that the DOPA-decarboxylase activity, which is low during the pupal stage in *Calliphora*, increases one day before adult emergence (Fig. 3.3) [45]. This increase in enzymic activity does not correlate with ecdysone titre, in contrast to the first peak at puparium formation.

Tanning of the cuticle in newly moulted insects (nymphal or adult) is controlled by a proteinaceous neurohormone, named bursicon [45,47]. Although this hormone has been studied mainly during emergence of adult blowflies and in the newly moulted nymphs of cockroaches (*P. americana*), it is not species-specific. When the head of a newly emerged blowfly is ligatured off, the remaining body does not harden or darken, but injection of haemolymph extracts containing bursicon activity causes both responses, whereas ecdysone will not. Conversely, bursicon is ineffective in moulting hormone bioassays. Unlike ecdysone-induced puparium formation, bursicon-induced tanning is not blocked by actinomycin or puromycin, and the hormone is thus presumably not affecting the synthesis of enzymes. It is supposed that bursicon controls tanning by increasing the permeability of haemocytes to tyrosine, thus facilitating its metabolism to tanning precursors (DOPA and dopamine), since the enzymes catalysing these transformations occur within the blood cells. Bursicon apparently functions by activating membrane adenyl cyclase to produce 3′, 5′-cyclic AMP (cyclic AMP), which is known to produce changes in membrane permeability in many systems. Involvement of cyclic AMP is supported by the demonstration of adenyl cyclase in insect haemocytes and the observations that cyclic AMP mimicks the action of bursicon by potentiating the darkening of ligated cockroach thoracic segments, as well as markedly enhancing the incorporation of radioactivity into the cuticle when injected simultaneously with tyrosine-[^{14}C] [48]. Although the basic tanning process in the post-ecdysial insect cuticle is essentially similar to the analogous transformation in fly puparium, the hormonal control mechanisms are quite different. Bursicon action, apparently mediated via cyclic AMP, promotes haemocyte permeability to tyrosine. In contrast, during puparial sclerotization, ecdysone promotes tanning precursor formation by inducing *de novo* synthesis of DOPA decarboxylase, as well as causing release of the puparium tanning factor (PTF) from the peripheral nerve endings (Fig. 2.10). In fact, the demonstration of the occurrence of PTF in a few insect species besides the higher Diptera suggests that this proteinaceous factor may also play some role (probably in concert with bursicon) in insects which tan after moulting [46].

It has been demonstrated in flies and cockroaches that bursicon is required not only for tanning of the exocuticle, but also for the post-moult deposition of endocuticle. Incorporation of leucine into the endocuticle is stimulated by the hormone, and in contrast to tanning, this process is inhibited by actinomycin and puromycin [47].

In addition to its role in tanning of the exocuticle and deposition of the endocuticle, bursicon has been claimed to control other physiological functions in newly ecdysed insects (see [49] for references). No doubt, unanswered questions regarding the target tissue and primary site(s) of action of bursicon will be greatly facilitated when its chemical identification is achieved, allowing use of the pure compound in investigations.

2.8 Microsomal oxidations and insecticide resistance

An area of much current practical importance concerns microsomal oxidation of insecticides, particularly in relation to insecticide resistance. Of the various insecticide detoxication mechanisms known, undoubtedly the systems most commonly encountered are the microsomal mixed function oxidases. The types of insecticide transformations catalysed by insect microsomes include [50]:

- Aromatic hydroxylation
- Hydroxylation of aliphatic hydrocarbons
- Epoxidation
- Oxidations of thioethers and α-phosphorothioates
- N-methyl hydroxylation
- O- and N-dealkylation

Developments in the understanding of microsomal oxidations in insects have largely followed progress with analogous mammalian systems. The liver microsomal hydroxylation system has been resolved into three components: cytochrome P-450, NADPH-cytochrome P-450 reductase (flavoprotein), and phosphatidylcholine. As in the mammalian system, insect microsomal mixed-function oxidases also involve cytochrome P-450 and NADPH-cytochrome P-450 reductase, but at present, information regarding the requirement of phospholipids is lacking [50]. In the liver system, the specificity for hydroxylation of different substrates apparently resides primarily in the cytochrome fraction, rather than in the reductase or lipid fraction. Multiple forms of cytochrome P-450 probably occur, which may also be the case for insect microsomes.

A number of compounds have remarkable activity as synergists for many insecticides in view of their potent microsomal mixed-function oxidase inhibitory activity, thus interfering with the metabolism of the insecticide molecules. The most important of these inhibitors are probably 1,3-benzodioxole (methylenedioxyphenyl)-containing compounds; the mechanism of inhibition is not clear [50].

We now turn to a consideration of insecticide resistance in insects. Insecticides are not mutagenic, and genetic variability for resistance already exists in natural populations which have not been exposed to insecticides. Several lines of evidence [51] support the proposal that development of resistance is biphasic. Phase I apparently involves selection of variants in the population which carry pre-adaptive genes for resistance, thus producing a resistance level commensurate with the gene pool orignally present. If it is assumed that resistance results in most cases from enzymic detoxication of the chemical, genetic selection will then produce an increase in enzymic activity. Phase II involves enhancement of resistance by induction of pre-existing detoxifying enzymes towards higher activity, thus resulting in more rapid degradation of the insecticide. The inducer is the insecticide itself, and this component of the total resistance depends upon the continued presence of the insecticide in the insect's environment, since its removal results in rapid reversion of resistance to the pre-existing genetic level.

Although many factors are involved in insect resistance to insecticides (e.g. reduced penetration of the toxicant, or decreased sensitivity of target enzymes to the chemical), the overwhelming predominant factor is the ability of the resistant insect to detoxify the chemical more rapidly and in larger quantities than its susceptible counterpart [see 52]. Both resistant and susceptible strains of insects essentially contain the same detoxifying enzymes, the differences being quantitative rather than qualitative.

The mechanism of induction of microsomal mixed-function oxidases by many insecticides, especially the chlorinated hydrocarbons such as DDT [1,1,1-trichloro-2,2-bis(p-chlorophenyl) ethane], is an area of much current interest. Although the mechanism of this induction is largely unknown, present evidence indicates that DDT acts at the nuclear level, and somehow stimulates transcription, resulting in increased synthesis of the mixed-function oxidase components, in particular cytochrome P-450, and in some instances NADPH-cytochrome P-450 reductase [50]. Information is not available regarding possible modification of the synthesis of microsomal phospholipids. In some cases, resistance to insecticides metabolized by microsomal mixed-function oxidases may be correlated with the levels of these enzymes. Similarly, cross-resistance to chemically unrelated insecticides may be due to common microsomal detoxifying mechanisms. Increased oxidative detoxication apparently accounts for the cross-resistance of some insects even to modern JH analogues (Chapter 5) [52].

The mechanism of induction of insect microsomal mixed-function oxidases by insecticides and the role of this process in insecticide resistance and cross-resistance await further clarification. Further elucidation of the mechanism of action of microsomal oxidases should provide a basis for the rational development of improved inhibitors for use as insecticide synergists.

2.9 Chemical control mechanisms: neurohormones and cyclic nucleotides

Neurohormones are produced within the nervous system, and serve as a link between neural communication and biochemical regulation [53]. They frequently co-ordinate responses in the animal with environmental change. Numerous environmental factors affect neurosecretory activity in insects. Whereas some, including temperature and possibly photoperiod, directly affect the neurosecretory cell, others stimulate sensory receptors to control the cell via activity within the central nervous system. The same basic mechanisms of synaptic transmission involving the same neurotransmitters have been found in all the vertebrate and invertebrate systems studied so far [54]. Although most developmental processes are controlled indirectly by the neurosecretory system through its action on the endocrine glands, a multitude of physiological processes are regulated directly by neurosecretions [47,55].

Purification and characterization of the active principle of neurosecretory material responsible for a particular control function is extremely difficult in view of the paucity of the glandular tissue concerned. Since the adipokinetic hormone of locust corpora cardiaca is the only insect neurohormone which is available in pure form [56], the picture is very complex and confused, and it is not known whether or not certain neurohormones may control more than one process. The active principles which have been partially characterized, generally appear to be polypeptides or proteins.

Since cyclic-AMP has been established as a second messenger in the action of many vertebrate hormones, it is convenient at this juncture merely to indicate an unusual feature regarding cyclic nucleotides in insects. It was originally demonstrated that whole crickets contained an unusually high concentration of 3′,5′-cyclic-GMP (cyclic-GMP) (2–4 nmol g^{-1}) in comparison with various rat tissues (0.01–0.07 nmol g^{-1}) and also contained more cyclic GMP than cyclic AMP, in contrast to vertebrate tissues. Functionally distinct cyclic AMP-dependent and cyclic GMP-dependent protein kinases occur and many lepidopteran tissues possess the latter almost exclusively [47,57]. Elucidation of the substrates of the latter kinase involved in the regulation of biological processes offers an exciting area for future investigation.

References

[1] House, H. L. (1974), in *The Physiology of Insecta, Second Edition,* (Ed.) M. Rockstein, Vol. V, pp. 63–117, Academic Press, New York.

[2] Dadd, R. H. (1970), in *Chemical Zoology,* (Ed.) M. Florkin and B. T. Scheer, Vol. V, pp. 117–145, Academic Press, New York.

[3] Treherne, J. E. (1967), *Ann. Rev. Entomol.*, **12**, 43–58.

[4] Sacktor, B. (1965), in *The Physiology of Insecta, First Edition*, (Ed.) M. Rockstein, Vol. II, pp. 483–580, Academic Press, New York.

[5] Sacktor, B. (1970), *Adv. Insect Physiol.*, 7, 267–347.

[6] Sacktor, B. (1974), in *The Physiology of Insecta, Second Edition*, (Ed.) M. Rockstein Vol. IV, pp. 271–353, Academic Press, New York.

[7] Hochachka, P. W. and Somero, G. N. (1973) *Strategies of Biochemical Adaptation*, W. B. Saunders Co., Philadelphia.

[8] Maruyama, K. (1974), in *The Physiology of Insecta, Second Edition*, (Ed.) M. Rockstein, Vol. IV, pp. 237–269, Academic Press, New York.

[9] Bailey, E. (1975), in *Insect Biochemistry and Function*, (Ed.) D. J. Candy and B. A. Kilby, pp. 90–176, Chapman and Hall, London.

[10] Kilby, B. A. (1965), in *Aspects of Insect Biochemistry* (Biochem. Soc. Symp. No.25), (Ed.) T. W. Goodwin, pp. 39–48, Academic Press, London.

[11] Wyatt, G. R. (1967), *Adv. Insect Physiol.*, **4**, 287–360.

[12] Sacktor, B. (1975), in *Insect Biochemistry and Function*, (Ed.) D. J. Candy and B. A. Kilby, pp. 3–88, Chapman and Hall, London.

[13] Friedman, S. (1970), in *Chemical Zoology,* (Ed.) M. Florkin and B. T. Scheer, Vol. V, pp. 167–197, Academic Press, New York.

[14] Crabtree, B. and Newsholme, E. A. (1975), in *Insect Muscle*, (Ed.) P. N. R. Usherwood, pp. 405–500, Academic Press, London.

[15] Bursell, E. (1967), *Comp. Biochem. Physiol.* **23**, 825–829.
[16] Hansford, R. G. and Sacktor, B. (1971), in *Chemical Zoology*, (Ed.) M. Florkin and B. T. Scheer, Vol. VI, pp. 213–247. Academic Press, New York.
[17] Chance, B. and Williams, G. R. (1956), *Adv. Enzymol.*, **17**, 65–134.
[18] Johnson, R. N. and Hansford, R. G. (1975), *Biochem. J.*, **146**, 527–535.
[19] Chen, P. S. (1971), *Biochemical Aspects of Insect Development*, S. Karger, London.
[20] Agrell, I. P. S. and Lundquist, A. M. (1973), in *The Physiology of Insecta*, *2nd Edition,* (Ed.) M. Rockstein, Vol. I, pp. 159–247, Academic Press, New York.
[21] Wyatt, G. R. (1968), in *Metamorphosis – a Problem in Developmental Biology*, (Ed.) W. Etkin and L. I. Gilbert, pp. 143–184, North-Holland Publishing Company, Amsterdam.
[22] Bücher, Th. (1965), in *Aspects of Insect Biochemistry*, (Biochem. Soc. Symp. No.25), (Ed.) T. W. Goodwin, pp. 15–28.
[23] Campbell, A. J. and Birt, L. M. (1975), *Insect Biochem.*, **5**, 113–117.
[24] Wagner, R. P. and Selander, R. K. (1974), *Ann. Rev. Entomol.*, **19**, 117–128.
[25] Asahina, E. (1969), *Adv. Insect Physiol.*, **6**, 1–49; Baust, J. G. and Miller, L. K. (1970), *J. Insect Physiol.*, **16**, 979–990.
[26] Harvey, W. R. and Haskell, J. A. (1966), *Adv. Insect Physiol.*, **3**, 133–205.
[27] Ziegler, R. and Wyatt, G. R. (1975), *Nature*, **254**, 622–623.
[28] Gilbert, L. I. and Chino, H. (1974), *J. Lipid Res.*, **15**, 439–456.
[29] Gilbert, L. I. and O'Connor, J. D. (1970), in *Chemical Zoology*, (Ed.) M. Florkin and B. T. Scheer, Vol. V, pp. 229–253, Academic Press, New York.
[30] Thompson, M. J., Kaplanis, J. N., Robbins, W. E. and Svoboda, J. A. (1973), in *Advances in Lipid Research*, (Ed.) R. Paoletti and D. Kritchevsky, Vol. 11, pp. 219–265, Academic Press, New York.
[31] Goodwin, T. W. (1971), in *Chemical Zoology,* (Ed.) M. Florkin and B. T. Scheer, Vol. VI, pp. 279–306, Academic Press, New York.
[32] Corrigan, J. J. (1970), in *Comparative Biochemistry of Nitrogen Metabolism.* (Ed.) J. W. Campbell, Vol. I, pp. 387–488, Academic Press, London.
[33] Linzen, B. (1974), *Adv. Insect Physiol.*, **10,** 117–246.
[34] Price, G. M. (1973), *Biol. Rev.*, **48**, 333–375.
[35] Cochran, D. G. (1975), in *Insect Biochemistry and Function*, (Ed.) D. J. Candy and B. A. Kilby, pp. 179–281, Chapman and Hall, London.
[36] Stobbart, R. H. and Shaw, J. (1974), in *The Physiology of Insecta*, *2nd Edition*, (Ed.) M. Rockstein, Vol. V, pp. 362–445, Academic Press, New York.
[37] Florkin, M. and Jeuniaux, C. (1974), in *The Physiology of Insecta*, *2nd Edition*, (Ed.) M. Rockstein, Vol. V, pp. 255–307, Academic Press, New York.
[38] Nappi, A. J. (1974), in *Contemporary Topics in Immunology*, Vol. 4, *Invertebrate Immunology*, (Ed.) E. I. Cooper, pp. 207–224, Plenum, New York. Also see several other chapters.
[39] Whitcomb, R. F., Shapiro, M. and Granados, R. R. (1974), in *The Physiology of Insecta*, *2nd Edition*, (Ed.) M. Rockstein, Vol. V, pp. 447–536, Academic Press, New York.
[40] Arnold, J. W. (1974), in *The Physiology of Insecta*, *2nd Edition*, (Ed.) M. Rockstein, Vol. V, pp. 201–254, Academic Press, New York.
[41] Garrod, D. R. (1973), *Cellular Development*, Chapman and Hall, London.
[42] Locke, M. (1974), in *The Physiology of Insecta*, *2nd Edition*, (Ed.) M. Rockstein, Vol. VI, pp. 123–213, Academic Press, New York.
[43] Lawrence, P. A. (1973), in *Developmental Systems: Insects*, (Ed.) C. J. Counce and C. H. Waddington, Vol. 2, pp. 157–209, Academic Press, London.
[44] Hackman, R. H. (1974), in *The Physiology of Insecta*, *2nd Edition*, (Ed.) M. Rockstein, Vol. VI, pp. 215–270, Academic Press, New York.
[45] Doane, W. W. (1973), in *Developmental Systems: Insects*, (Ed.) C. J. Counce and C. H. Waddington, Vol. 2, pp. 291–497, Academic Press, London.
[46] Sivasubramanian, P., Friedman, S. and Fraenkel, G. (1974), *Biol. Bull.*, **147**, 163–185.
[47] Wyatt, G. R. (1972), in *Biochemical Actions of Hormones*, (Ed.) G. Litwack, Vol. II, pp. 385–490, Academic Press, New York.
[48] Vandenberg, R. D. and Mills, R. R. (1975), *J. Insect Physiol.*, **21**, 221–229.
[49] Reynolds, S. E. (1976), *J. Insect Physiol.*, **22**, 529–534.
[50] Agosin, M. and Perry, A. S. (1974), in *The Physiology of Insecta*, *2nd Edition*, (Ed.) M. Rockstein, Vol. V, pp. 537–596, Academic Press, New York.

[51] Perry, A. S. and Agosin, M. (1974), in *The Physiology of Insecta, 2nd Edition*, (Ed.) M. Rockstein, Vol. VI, pp. 3–121, Academic Press, New York.
[52] Plapp, F. W. (1976), *Ann. Rev. Entomol.*, **21**, 179–197.
[53] Treherne, J. E., Ed. (1974), *Insect Neurobiology*, North-Holland, Amsterdam.
[54] Lunt, G. G. (1975), in *Insect Biochemistry and Function,* (Ed.) D. J. Candy and B. A. Kilby, pp. 285–306, Chapman and Hall, London.
[55] Goldsworthy, G. J. and Mordue, W. (1974), *J. Endocr.*, **60**, 529–558.
[56] Stone, J. V., Mordue, W., Batley, K. E. and Morris, H. R. (1976), *Nature, (Lond.),* **263**, 207–211.
[57] Nishiyama, K., Katakami, H., Yamamura, H., Takai, Y., Shimomura, R. and Nishizuka, Y. (1975), *J. Biol. Chem.*, **250**, 1297–1300.

Further reading

General review articles have been cited e.g. [9] and [12].

Gilmour, D. (1965), *The Metabolism of Insects*, Oliver and Boyd, Edinburgh.
'A short account of the earlier work on insect intermediary metabolism'.

Wigglesworth, V. B. (1972), *The Principles of Insect Physiology, 7th Edition*, Chapman and Hall, London.
'A comprehensive reference book of insect physiology'.

The following are comprehensive, specialized texts

Usherwood, P. N. R., Ed. (1975), *Insect Muscle,* Academic Press, London.

Hepburn, H. R., Ed. (1976), *The Insect Integument*, Elsevier, Amsterdam.

3 Hormonal control of development

3.1 Endocrine control of moulting and metamorphosis

The 'classical scheme' for the hormonal control of moulting and metamorphosis involves an endocrine system consisting of the brain and associated glands innervated by the brain (corpora allata and corpora cardiaca), together with the paired prothoracic glands in the prothorax [1,2]. In response to various stimuli (e.g. temperature change, photoperiod, distention of the gut after feeding) neurosecretory cells in the brain produce brain hormone (ecdysiotropin, prothoracicotropic hormone), which enters the haemolymph, in many cases via the corpora cardiaca. Brain hormone then stimulates the prothoracic glands to synthesize and/or release insect moulting hormone (MH). Under direct neural control from the brain, the corpora allata secrete juvenile hormone (JH). Moulting hormone initiates moulting, whereas JH determines the morphological outcome of the moult by suppressing adult differentiation in favour of larval structures (Fig. 3.1). In hemimetabola, a decline in JH titre leads to the nymphal-adult moult, whereas it has been widely accepted that in holometabola, the larval-larval, larval-pupal, and pupal-adult moults occur in the presence of high, intermediate, and negligible titres of JH, respectively. Recent reports indicate that hormonal control of insect development is more complex, and this scheme may need modification.

In addition to its effects on the epidermal cells, JH also influences the development of the internal organs, preventing the changes which usually accompany metamorphosis e.g. the evagination of imaginal discs and maturation of the digestive tract. The corpora allata cease to control metamorphosis at pupation, but in many species (not all) these glands resume activity in the adult, when JH controls entirely distinct processes connected with reproductive maturation and activity. In most species, during the adult moult the prothoracic glands undergo degeneration apparently programmed by the absence of JH [1].

Brain hormone has proved particularly difficult to purify, but preliminary studies suggest that it is a polypeptide. Activation of the prothoracic glands by brain hormone may be mediated via adenyl cyclase activation [3].

3.2 Determination and differentiation

Cell differentiation is the development in higher organisms of different cell types arising from a single cell type, the zygote. This topic has been considered elsewhere in this series [4]. Differentiation essentially commences with the process of *determination*, which programmes a population of cells for one particular developmental pathway to the virtual exclusion of others [5]. The developmental programmes represent sets of genes controlled in a closely co-ordinated manner. Two major categories of determinative processes have been recognized: *embryonic induction* involves cell or tissue interactions via diffusible inducing substances,

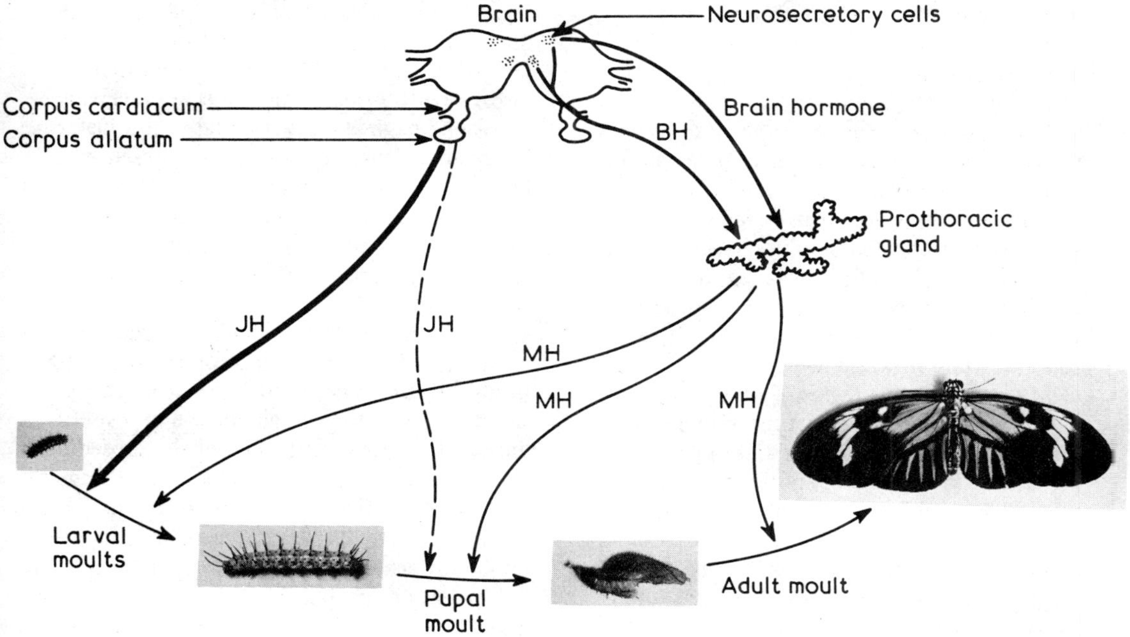

Fig. 3.1 Classical scheme of the hormonal control of moulting and metamorphosis in a holometabolous insect.

whereas *cytoplasmic segregation* is dependent upon factors intrinsic to the differentiating cells.

In holometabolous larvae, groups of cells **(imaginal discs)** arise as outgrowths of the embryonic hypodermis, and give rise during metamorphosis to the entire adult (imaginal) integument as well as many internal structures; different parts of the adult body are derived from different discs. During the larval stages, the discs grow considerably by cell division, but onset of metamorphosis during the pupal phase of development is characterized by a switch (triggered by a high MH titre in the absence of JH) from disc cell multiplication to terminal cell differentiation, leading to formation of the corresponding adult structures.

The imaginal discs of *Drosophila* and other holometabolous insects provide a particularly suitable model system for developmental studies, especially since Hadorn [6] has developed a method whereby disc cells can be cultured permanently (for several years) in a determined state in the abdomen of adult female flies whose haemolymph serves as a culture medium. Differentiation of the disc cells can be triggered experimentally by exposure to MH either *in vitro,* or *in vivo* in mature larvae [7,8]. Therefore, in imaginal discs, the determination for a particular genetic developmental programme is clearly separated from the execution of that programme.

Drosophila is unique among multi-cellular organisms because genetic techniques can be used for analysis of the regulation of developmental pathways. These techniques allow studies of cell interaction, pattern formation [7] and determination.

Although imaginal discs are not apparent as distinct groups of cells until late embryonic stages, the basic process of determination for adult structures occur much earlier. Determination apparently proceeds in a stepwise manner by restricting progressively the developmental capacities of disc cells. By the end of larval development, the cells may be committed to a very narrow pathway of development i.e. to form a certain region, but not yet a particular cell type within this region.

During adult differentiation of imaginal discs, MH (in the absence of JH) induces the expression of those genes in the disc cells, that are specified by their determination. However, hormones are not the only parameters involved in initiating adult differentiation; an imaginal disc must be '*competent*' to respond to the hormonal stimuli. The factors involved in the acquisition of competence and, therefore, the ability of the cells to switch from multiplication to imaginal differentiation are unknown; there may be a requirement for a minimum number of cell divisions in the disc primordium.

A specific state of determination can be propagated *in vivo* over numerous cell generations, and is, consequently, to some extent stable. Since determination is apparently carried by the individual cell rather than by the organ as a whole, the mechanism for propagation of the determined state is designated as *cell heredity.* A crucial major problem is the identification of the *carrier of determination,* i.e. the cell component which carries the information for future differentiation. The stability of the determined state is limited, since in the cultured discs occasional changes of determination (transdetermination) occur in clones derived from determined cells. For example, genital discs cultured *in vivo* occasionally yield not only genital structures but also antennal and leg structures which in normal development are formed by other discs. Most transdeterminations are reversible.

When imaginal discs from late third instar larvae are cultured *in vitro* in the presence of MH, they undergo morphogenesis. This involves initially evagination, which is followed by synthesis and deposition of cuticular proteins and differentiation of normal adult cuticular structures [9].

Differentiated cells are characterized primarily by the structural proteins and sometimes the enzyme proteins they produce e.g. silk proteins of the silk gland, and myosin of muscle. Consequently, much of the subsequent discussion of hormonal control of development and differentiation will deal with the mechanism of hormone-directed synthesis of specific protein. Synthesis of a comparatively large amount of a particular protein by differentiated cells at a certain time in their development is accomplished in many cases by producing a stable mRNA, thus amplifying the effect of the specific gene during protein synthesis. Undoubtedly, immunological techniques will play an increasing role in investigations into the biochemical mechanisms of development and differentiation (*see* for example [10]).

3.3 Moulting hormones

3.3.1 Chemical nature and occurrence

Several insect moulting hormones have been identified [11,12] (Fig. 3.2); the name 'ecdysone', originally assigned to a specific compound (α-ecdysone), is now generally used for all steroidal compounds which exhibit activity in insect-moulting hormone bioassays. The original isolation of α-ecdysone employed the *Calliphora* bioassay. This involves ligaturing mature larvae of *Calliphora erythrocephala,* thus preventing hormone from the anterior portion from diffusing into the posterior region. Injection of hormone into the posterior region of animals

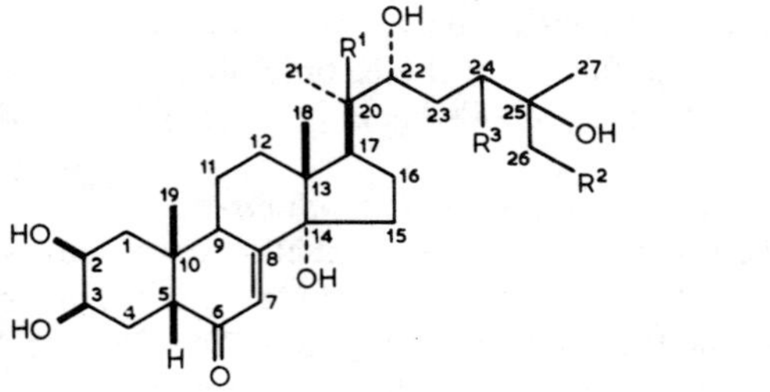

α-ecdysone (ecdysone)	: $R^1 = R^2 = R^3 = H$
β-ecdysone Ecdysterone 20-hydroxyecdysone Crustecdysone	: $R^1 = OH, R^2 = R^3 = H$
20, 26-dihydroxyecdysone	: $R^1 = R^2 = OH, R^3 = H$
26-hydroxyecdysone	: $R^1 = R^3 = H, R^2 = OH$
Makisterone A	: $R^1 = OH, R^2 = H, R^3 = CH_3$

Fig. 3.2 Structures of ecdysones isolated from insects.

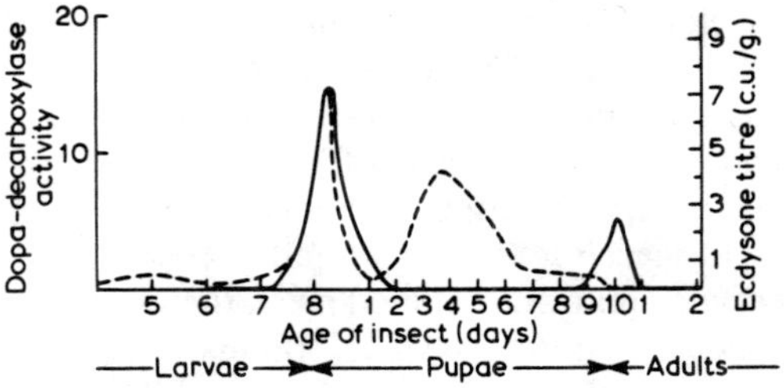

Fig. 3.3 Ecdysone titre and dopa-decarboxylase activity during development of *Calliphora erythrocephala.* (– – – – –) Ecdysone titre in *Calliphora* units/g; the *Calliphora* unit (C.U.) of biological activity is defined as the amount of material necessary to produce 50–70% pupation in ligated abdomens of the blowfly. In most cases this effect is given by ≃0.01 μg α-ecdysone. (———) DOPA decarboxylase activity (as percentage transformation of DOPA to dopamine). Redrawn from [13].

which after 24 h have only pupated in the anterior part induces formation of a puparium which is noticeable as a darkening and hardening of the cuticle.

Bioassay and, more recently, radioimmunoassay, gas-liquid chromatographic, and high-performance liquid chromatographic techniques have been used to determine the MH titre during post-embryonic development. Although the titre curves for various holometabolous insects differ quantitatively, the general features are remarkably similar. A prominent peak of MH is generally noted at the beginning of larval-pupal metamorphosis, and a subsequent peak may be related to adult development during the pupal stage (Fig. 3.3).

β-ecdysone is generally the predominant MH during post-embryonic development, and is usually accompanied by smaller amounts of α-ecdysone [11,12]. In the pupae (i.e. during pupal-adult development) of tobacco hornworm, *Manduca sexta,* β-ecdysone is accompanied not only by smaller amounts of α-ecdysone but also 20,26-dihydroxyecdysone [14].

Recent results have led to the fascinating realization that ecdysones may function not only in moulting, but also during the embryonic and adult stages of the insect life cycle. During pupal-adult development in the silkmoth, *Bombyx mori,* a single peak in MH titre occurs on the second day after pupal ecdysis in males, whereas females exhibit a distinct second peak on the fifth day when the ovary grows rapidly [15]. This second peak, apparently due to the accumulation of MH in the ovary, may be eliminated by ovariectomy during the larval stage. The hormone may be involved in ovarian maturation and, in fact, MH occurs in the maturing oöcytes [16] but it is not clear whether the hormone is synthesized by the developing ovaries or transferred to them from the prothoracic glands. The peak of hormonal activity observed in both sexes at the early period of pupal-adult development apparently controls adult differentiation. This is also the only peak observed during pupal-adult development of females of other species of insects whose ovaries develop after adult emergence.

In the adult female mosquito, *Aedes aegypti,* following a blood meal, the ovary produces a hormone which stimulates vitellogenin (yolk protein) synthesis in the fat body. The ovarian hormone is apparently α-ecdysone which is hydroxylated (e.g. in the fat body) to β-ecdysone, the hormone actually stimulating the synthesis of vitellogenin [17]. Moulting hormone has also been implicated in reproduction in other species of insects. For example, in *Locusta,* the MH titre of the ovary increases during chorion (proteinaceous egg shell) formation in each cycle of oöcyte maturation [18]. It is conceivable that MH could be involved in regulation of the sequential synthesis of chorion proteins.

A considerable increase in MH activity has been observed in the eggs of several species of insect during embryogenesis. It is significant that 26-hydroxyecdysone is the major hormone in *Manduca sexta* eggs during late embryonic development, although α-ecdysone, β-ecdysone, and 20,26-dihydroxyecdysone are the major hormones during the pupal state [19]. The function of the embryonic hormones is unknown. Since many insects moult once or twice toward the end of embryogenesis, the embryonic hormones may function in the control of these processes. It is apparent that the various ecdysones not only control post-embryonic development, but also function in adult reproduction and probably in embryogenesis, and are, thus, involved in the regulation of the entire life cycle of, at least, some insects.

Ecdysones also occur and exhibit biological activity in arthropods besides insects. In addition, approximately forty closely related compounds possessing MH activity have been isolated from the plant kingdom, but they are not by any means universally distributed. The significance of these compounds in plants is something of an enigma [11,12].

3.3.2 Biosynthesis and metabolism

Ecdysones are formed from cholesterol, but elucidation of the intermediate steps has proved recalcitrant, largely owing to the low level of hormone production. Investigation of the metabolism of radioactively labelled putative intermediates indicates that some elaboration of the ring system must precede side chain hydroxylation, and that more than one hydroxylation sequence may exist to some extent even within a single insect species, particularly at different developmental stages [1,14]. A plausible scheme for the biosynthesis of ecdysones in *Calliphora* is shown in Fig. 3.4 [20]; hydroxylation of the side chain may precede elaboration of the complete nuclear structure in some species.

According to the 'classical scheme' for the hormonal control of insect post-embryonic development, ecdysones should be synthesized

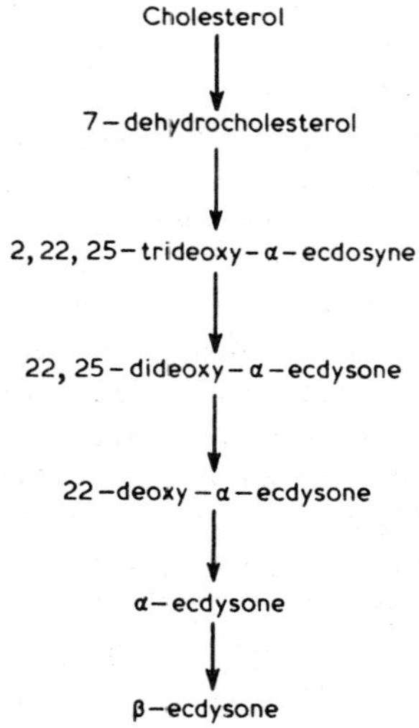

Fig. 3.4 Plausible biosynthetic scheme of ecdysones in *Calliphora*.

in the prothoracic glands. The fact that prothoracic glands cultured *in vitro* produce exclusively α-ecdysone with no detectable β-ecdysone provides direct evidence for this. Many tissues, including the fat body can rapidly convert α-ecdysone (primarily a prohormone see below) into β-ecdysone (the principal hormone) whereas others (e.g. salivary glands and imaginal discs) cannot effect this transformation [21]. The latter tissues must effect uptake of β-ecdysone (produced by other organs) from the haemolymph. Numerous observations suggest the existence of extraprothoracic gland site(s) of production of ecdysones in larvae, but this is quantitatively much less important than that by the prothoracic glands [1,22].

Ecdysones, following their synthesis, are transported in the haemolymph possibly via binding proteins to target tissues, where they exert their action. Some reports describe weak, reversible binding of α-ecdysone to haemolymph proteins, but in other cases there is said to be no evidence for such specific binding [21]. In view of the appreciable water solubility of ecdysones, a haemolymph carrier protein might not be essential.

The rapid changes in MH titres during development necessitate rigorous control mechanisms in the biosynthesis-inactivation-excretion processes. Inactivation of ecdysones occurs principally by formation of sulphate conjugates and, probably to a lesser extent, glycoside derivatives [14,19]. Ecdysones can also apparently be converted reversibly into 3-dehydro products, which may in turn be inactivated by sulphation. Excretion of ecdysones in the free and conjugate forms is reduced drastically at the time corresponding to a high hormone titre, suggesting the existence of a regulatory mechanism [23]. The significance of the formation of the hormonally less active 3α-epimers of ecdysones in midgut of *Manduca sexta* is unknown, but may be involved in metabolic control of MH titre [19].

3.3.3 Effects of ecdysones

There has been a deluge of reports on the effects of ecdysones and juvenile hormones *in vivo,* and also upon isolated tissues *in vitro* [1,2,24]. However, there is always a possibility that the effects *in vivo* are indirect.

α-Ecdysone may have direct action upon target tissues in some species of insects e.g. in *Chironomus* salivary glands, α-ecdysone at very high concentrations elicits puffing at specific chromosomal loci that are not affected by β-ecdysone. It appears, however, that α-ecdysone is primarily a prohormone, being transformed by C–20 hydroxylation into β-ecdysone. For example, imaginal disc or polytene chromosome systems *in vitro* usually require α-ecdysone at concentrations 100–300 times greater than β-ecdysone to elicit equivalent responses. In this respect, it is significant that the ratio of β:α-ecdysone changes during the final larval or nymphal instar, becoming especially high during the peak titre of total moulting hormone activity. Specific functions of α-ecdysone in certain tissues are not precluded.

A major event initiated by MH is the moult cycle, and, in the case of Diptera, puparium formation. These essentially epidermal events provide several experimentally amenable systems for investigation of hormone action. There is no doubt that ecdysone acts directly upon epidermal structures *in vitro.* The moult cycle consists of a definite sequence of events: apolysis, moulting fluid production, initiation of cuticle formation, moulting fluid activation, exocuticle secretion, ecdysis, tanning, and endocuticle deposition. Similarly, the numerous events involved in puparium formation culminate in tanning of the last larval cuticle. Both these series of events require careful co-ordination, and the question arises as to whether ecdysone merely serves as a trigger, or sustains and co-ordinates the entire sequence of actions. Some intermediate situation is quite likely.

Locke has suggested that the developmental sequence is linked at the genetic level in the epidermal cells and that ecdysone functions as the initial stimulus with other stimuli (both extrinsic and intrinsic) acting at later stages to control quantitatively the expression of individual phases of the sequence [25]. The occurrence of low levels of ecdysone during the entire last larval instar and the fact that even the peak ecdysone titre is generally considerably less than the amount required to obtain effects by exogenous application suggest that subliminal effects may be exerted by the endogenous hormone prior to the peak titre. High doses of ecdysones often cause such rapid moulting that some normal events are by-passed with production of abnormalities in the resulting insects. Whereas some processes in the moult cycle are under immediate control of ecdysone and may be initiated quickly by high doses, the normal developmental sequence, involving an ordered progression of effects is dependent upon the continued presence of lower levels of the hormone [24,26].

The imaginal discs of immature insects are a system of epidermal origin, which has been utilized extensively for studies on ecdysone action [7]. Whether ecdysone is required for growth of imaginal discs is a controversial issue [2]. Different ecdysones in the absence of JH promote various degrees of morphogenesis and differentiation in Dipteran and Lepidopteran discs cultured *in vitro*, the MH being required as a sustained stimulus of metamorphosis. Studies with imaginal discs from *Galleria* (waxmouth) *in vitro* suggest that α- and β-ecdysone may have separate functions in initiating imaginal development [27]. Contradictory results have been reported for the relative activities of α- and β-ecdysones for differentiation of *Drosophila* imaginal discs *in vitro*, but it now seems clear that β-ecdysone is the principal hormone responsible for differentiation in this system [28,29].

Although the action of ecdysones on the epidermis has been emphasized, other tissues undergo either growth and differentiation or histolysis in ecdysone-initiated moulting and metamorphosis, e.g. ecdysones exert direct effects on salivary glands, gonads, and intestinal tissue.

Exogenous application of ecdysone ends diapause in some species of insects, although developmental arrest is the result of a lack of ecdysone in by no means all cases [2]. When administered simultaneously with β-ecdysone cyclic GMP enhances, but cyclic AMP inhibits the diapause-breaking action of the MH in *Mamestra configurata* pupae; cyclic GMP cannot by itself break the pupal diapause [30]. The mechanism of these opposing regulatory influences is obscure.

3.4 Juvenile hormones

The most important effects of JH include its morphogenetic role in maintaining the larval (or nymphal) state in immature insects, and also regulation of entirely distinct processes connected with reproductive maturation and activity in adults.

3.4.1 Chemical nature and occurrence

The early isolations of JH from insects were monitored by sensitive bioassay procedures, based upon the ability of the hormone to 'prevent' the differentiation of adult structures. Local administration of JH-active extracts to pupae of holometabola or to last instar nymphs of hemimetabola causes the pupal or nymphal epidermal cells to synthesize a second pupal (or nymphal) cuticle, whereas the adjoining uninfluenced cells undergo adult ecdysis [24].

The original isolation and structural elucidation of the major juvenile hormone (JH–I; Fig. 3.5) from *Hyalophora cecropia* (Lepidoptera) was achieved by Röller's group [31]; JH–I was later shown to be accompanied by

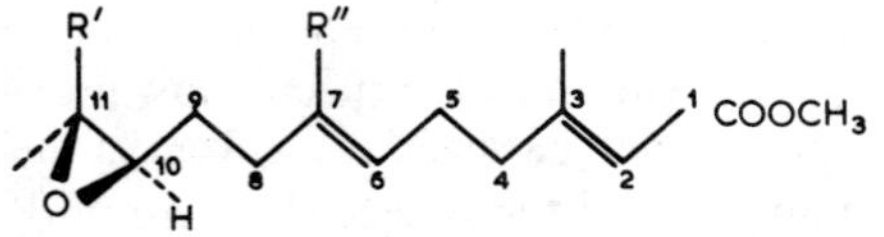

JH – I : $R' = R'' = C_2H_5$

[Methyl (2*E*,6*E*) –(10*R*, 11*S*) – 10,11 – epoxy –7– ethyl –3, 11 – dimethyl –2, 6 – tridecadienoate]

JH – II : $R' = C_2H_5$, $R'' = CH_3$

JH – III : $R' = R'' = CH_3$

Fig. 3.5 Structures of juvenile hormones.

smaller amounts of a lower homologue (JH-II) [32]. The difficulty in isolating and identifying minute amounts of JH from whole insects has been overcome in various ways. For example, larger amounts of hormone have been obtained from media in which corpora allata have been cultured *in vitro.* This approach allowed the identification of the hormones of the tobacco hornworm (*M. sexta*) as JH-II together with the lower homologue, JH-III (Fig. 3.5) [33]. However, it is always possible that the products of JH biosynthesis *in vitro* may not reflect qualitatively or quantitatively the juvenile hormones normally occurring *in vivo.* This problem is not encountered in a new sensitive method involving high pressure liquid chromatography followed by gas-liquid chromatography (g.l.c.)-chemical ionization mass spectrometry for isolation and identification of JH from haemolymph [34]. This technique allowed the demonstration that in the hemimetabolous cockroach, *Nauphoeta cinerea,* not only the total JH titre, but also the relative concentrations of the three known juvenile hormones change during development, thus suggesting the possibility of some degree of diversification in function between the different hormones. JH-III is apparently the predominant JH in a number of adult female insects, where the hormone exerts gonadotropic effects. It may be significant that the relative activities of JH-I, II and III vary tremendously between insects [1].

Although there is little information on the changes in titres of JH during insect development, generally the level is high in larvae, low or absent during pupation and adult development, but high in the mature adult [24]. This pattern is consistent with the major roles of JH in maintaining the larval state and in regulating reproductive maturation and activity in many adult insects.

3.4.2 Biosynthesis, transport and metabolism

A possible pathway of JH-II biosynthesis has been suggested from results of experiments on the incorporation of radioactive acetate, mevalonate, and propionate into juvenile hormones (Fig. 3.6) [35]. According to this scheme, JH biosynthesis proceeds via a basic terpenoid route in which the isoprenoid units arise from mevalonate, whereas the homo-isoprenoid units arise from one propionate and two acetate residues via homomevalonate. JH-I arises from two homo-isoprenoid and one normal isoprenoid units [36]. A striking feature of JH-I and II biosynthesis is the remarkable enzymic specificity in the assembly of only limited arrangements of the isoprenoid and homo-isoprenoid units. The sequence of epoxidation at C-10–C-11, and of methylation at C-1 by S-adenosyl methionine during the terminal stages of hormone biosynthesis remains uncertain; two sequences may exist. The mechanism controlling JH synthesis by the corpora allata is also not well understood.

Following synthesis in the corpora allata, JH is secreted into the haemolymph in which it is transported to the target cells, where it fulfils its function(s). At the appropriate time, it undergoes inactivation via two principal pathways (Fig. 3.7) [37]. The carboxylesterase and epoxide hydrase enzymes involved occur in a number of tissues, but the latter microsomal enzyme is absent from the haemolymph.

The lipid-soluble JH could reasonably be expected to be transported in the haemolymph

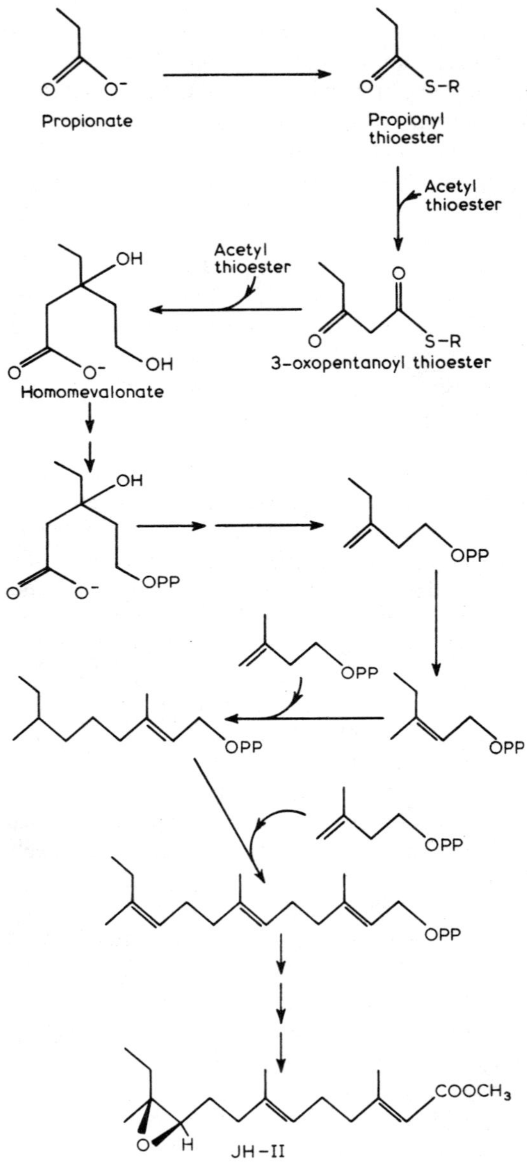

Fig. 3.6 Possible biosynthetic pathway of JH-II. Redrawn from [35].

in association with a hydrophilic carrier, probably a protein. In fact, a specific low mol.wt. binding protein has been recognized in the haemolymph of the tobacco hornworm, *Manduca sexta* [38], and also some other insects. Besides functioning in the transport of JH in the haemolymph, such proteins apparently also have a role in protecting the hormone from degradative enzymes (see below), as well as possibly being involved in cellular recognition of hormone at the target tissues [39]. JH may also be associated with a specific component of the haemolymph high density lipoprotein (HDL) in some species of insects [21].

Of importance in understanding the role of JH in the control of metamorphosis is the finding that haemolymph from final (fifth) instar larvae of *M. sexta* contains two families of esterases distinguishable by their reactivity with diisopropylphosphorofluoridate (DFP) [39]. One group consists of general esterases, which can hydrolyse free JH but not JH complexed to the binding protein, and are completely inhibited by low concentrations of DFP. The other group (JH-specific esterases) is relatively DFP resistant and has little detectable general esterase activity but can hydrolyse free JH or JH complexed to the binding protein. The general esterases are present throughout the larval stages, whereas the JH-specific esterases are barely detectable until late in the fifth instar (just prior to metamorphosis), when they appear rapidly at a high concentration, causing hydrolysis of JH complexed to the carrier protein. Therefore, by lowering the JH titre at the appropriate time, the JH esterases are important factors in the control of metamorphosis.

3.4.3 Effects of juvenile hormone

It will only be possible in this treatise to pay a mere cursory obeisance to the effects of JH; the topic has been reviewed [1,2,26]. An

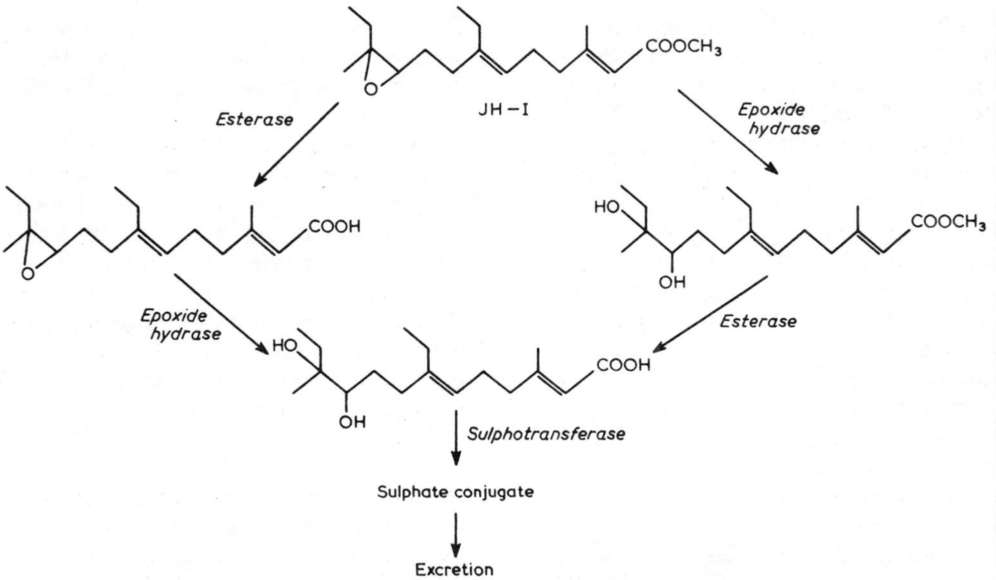

Fig. 3.7 Major metabolic pathways of juvenile hormone.

important target tissue is the epidermis, upon which JH (like MH) acts directly. The effectiveness of JH in influencing the type of cuticle to be produced ceases long before the beginning of cuticle secretion. It is clear that the developmental fate of the epidermis in various body regions is determined at different times [40].

According to the widely accepted 'classical' scheme, control of insect metamorphosis involves ecdysone in causing a moult, with the *titre* of JH directing the morphological outcome of the moult (i.e. the type of cuticle secreted); high titres of JH result in larval (or nymphal) moults, low titres yield larval-pupal transformation, whereas in its absence hemimetabolous nymphs and holometabolous pupae moult into adults. Thus, in larvae or pupae moulting in the presence of high titres of JH, the epidermal cells secrete a new cuticle, which would retain the characteristics of the preceding stage. This would constitute a *status quo* action of JH. An essential attribute of this 'classical concept' is that the pupal or adult epidermal cells can apparently, in some cases, revert to secreting a larval cuticle in the presence of high JH titres. The hormone should, thus, cause 'reversal of metamorphosis'.

The 'classical' concept has been retested recently by investigating the developmental fates of larval, pupal, and adult epidermal cells in a single animal under identical JH concentration in the haemolymph [41]. Such animals (heterochronic) can be produced by application of analogues of MH and of JH. According to the 'classical' concept, the different cells in heterochronic animals should develop synchronously into a homochronic population of either larval, pupal, or adult epidermal cells according to the JH titre. If this is correct, heterochronic character should not exist in more than one instar. However, it was observed that the heterochronic character of the tissues never developed into a uniform population of homomorphic epidermal cells. Rather, in the presence of an effective JH titre, the heterochronic pattern was fully preserved and carried on into the next instar. Furthermore, in the absence of effective JH titres, the ontogenetically different tissues (i.e. of different developmental age), such as larval and pupal epidermal cells, simultaneously underwent their respective larval-pupal or pupal-adult morphogenesis in the same hormonal milieu. Although the exact implications of these results are uncertain, they suggest that not only is the JH titre an important factor in determination of the type of morphogenic changes, but also the extant, previously attained degree of ontogenetic structural differentiation. Whereas JH can temporarily and reversibly inhibit morphogenetic progress at different ontogenetic levels, there are conflicting reports as to whether it can cause reversal of metamorphosis at any of these levels.

During postembryonic development, JH usually (but not invariably) acts as a modulator of MH action. A negligible titre of JH is not the only prerequisite for establishment of the pupal programme; a sub-threshold level of MH is also necessary. Activation of this programme requires a threshold level of MH.

Many reports suggest that the action of JH is connected in some way with DNA synthesis. A problem in understanding the action of JH is whether or not the same epidermal cell can secrete larval, pupal, and adult cuticle without being reprogrammed by DNA replication before metamorphosis. Although in many cases JH is effective in preventing metamorphosis only when administered prior to DNA synthesis, there is conclusive evidence that JH does not block DNA synthesis. In the case of vertebrate systems, at least one cycle of DNA synthesis may be critical for commitment to a new course of differentiation or transformation of cellular function. By analogy, it would be expected that if DNA synthesis had occurred, cells would be refractory to JH. Although there is much evidence that reprogramming by JH requires DNA replication, the occurrence of instances where JH affects cells after the completion of DNA synthesis provides a crucial challenge to this notion.

During the final larval instar of the tobacco hornworm, JH inhibits the secretion of prothoracicotropic hormone (BH), thus obviating another larval moult [42]. When the larvae attain a certain weight, the corpora allata cease to secrete JH, which disappears from the haemolymph, a situation that renders the brain capable of releasing BH. Thus, the turning-off of the corpora allata, by an unknown mechanism, is not only a factor leading to the formation of pupal characters when MH causes the succeeding moult, but also triggers the endocrine events which lead to the moult itself. In this connection, it is interesting that a prothoracicotropic action (i.e. stimulation of the prothoracic glands) by JH has also been demonstrated. For a full discussion of corpora allata-prothoracic gland interactions, see [1].

Another morphogenetic action of JH is its ability to interfere with embryonic development when applied to the egg or the mother. In certain insects, onset of adult diapause is due to lack of JH; neurosecretory cells of the brain, influenced by an environmental cue, control the activity of the corpora allata.

In the adult insect, JH controls the functional activity of the accessory reproductive glands in both sexes and in females of many species it regulates both the synthesis of yolk proteins (vitellogenins) in the fat body and their uptake from the haemolymph by the maturing oöcytes. JH is also probably involved in lipid metabolism in some adult insects, presumably acting so that lipid stored in the fat body can be shunted to the developing oöcytes to function as a substrate for embryogenesis. For a full discussion of the role of JH in the control of reproductive metabolism see [1,24]. Thus, it is clear that the functions of JH in adults are entirely different from its role in controlling morphogenesis in immature insects. It is thus possible to use vitellogenin synthesis in the adult, or epidermal development in the immature insect as excellent model systems in which to investigate at the molecular level the manner in which one type of hormone can function in entirely different ways at different stages of development.

JH analogues have been widely used in biological experiments, but assertions that such compounds act by synergising subthreshold levels of endogenous JH rather than possessing intrinsic hormonal activity are probably not tenable [43].

3.5 Biochemical mode of action of ecdysones

Intracellular receptor proteins which bind steroid hormones occur in vertebrate target tissues, but are virtually absent from non-target organs. It is envisaged that the hormone enters the target cell, becomes bound to the cytoplasmic receptor, which perhaps then undergoes a transition into a form capable of entering the nucleus, where the hormone-receptor complex presumably exerts its action leading to specific gene activation. Analogous MH receptor proteins occur in insect target tissues, thus making the operation of a similar scheme plausible. In contrast, other hypotheses suggest that the primary action of MH may be extra-nuclear. It is noteworthy that most basic ideas behind current theories of the molecular mode of action of steroid hormones were derived originally from work on ecdysones.

It is necessary first of all to consider an almost distinctive insect experimental system (polytene chromosomes) which has been particularly instrumental in investigations on the mode of action of insect developmental hormones at the molecular level [44,45].

3.5.1. Polytene chromosomes and puffing

The giant (polytene) chromosomes in various tissues of certain Diptera (and some Collembola) provide a developmental system in which evidence of gene activity can be observed under the microscope. Polytene chromosomes are produced when each chromosomal element of a diploid nucleus duplicates successively without segregation of the daughter chromosomes, which remain together as a giant multi-stranded chromosome. For example, salivary gland cells of *Drosophila* contain 500–1000 copies of each chromatid. There is variation in the degree of coiling of the DNA and its associated histones along chromosomes, regions of high concentrations being known as 'chromomeres'. The pattern of chromomeres is highly specific for each chromosome, and in polytene chromosomes homologous chromomeres are in precise alignment, furnishing a characteristic band pattern [4].

The correlation of certain mutant loci with particular polytene chromosome bands, led to the notion that each band may correspond to a single classical gene. This view is probably an over-simplification since there is far more DNA per band than is required to code for a single protein [46].

At certain times, many polytene chromosomal bands produce 'puffs' i.e. become swollen. This reversible structural modification involves some uncoiling of the chromosome fibres to allow synthesis of specific RNA species, which, together with acidic proteins, accumulate at the particular band site. Only a proportion of polytene chromosomal bands that are active in RNA synthesis become visibly puffed. The few exceptionally large puffs of Chironomids are called 'Balbiani rings'.

Puffing changes show both tissue and developmental (time) specificity. The notion that puffs are a visual manifestation of gene activity is now strongly supported by experimental evidence. For example, artificial conditions, such as heat shock, which change the puffing pattern of salivary gland chromosomes, also change, in parallel, the pattern of protein synthesis. The newly synthesized RNA, produced in response to heat shock, not only hybridises specifically to the puff sites induced by that treatment, but is also translated *in vitro* into the same proteins as are produced by heat-shocked tissues [47].

Most puffs appear at specific stages during development, remain active for a time, and then regress. The observation that periods of moulting and the early stages of metamorphosis are associated with intense puffing activity led Clever and Karlson to implicate MH in the control of puffing. They showed that injection of α-ecdysone into early final instar larvae of the midge, *Chironomus tentans*, initiated prematurely the changes in salivary gland puffing activity that would normally occur prior to the larval-pupal moult. In this case, two 'early' puffs appeared within 15–60 minutes of hormone administration, followed by various 'late' puffs between 5 and 72 hours. Ecdysone, being a hormone which causes differential gene activation, induces different puffing changes in different tissues. How is the correct temporal sequence of induction of a large number of puffs during development controlled? Hormonal induction of early puffs in salivary glands of *Chironomus tentans* is independent of protein synthesis, whereas the appearance of the late puffs requires such synthesis, suggesting that the early puffs might control the activity of the late puffs.

Studies on the control of puffing activity by MH are instrumental in understanding the mechanisms not only of hormone action, but also of control of a temporal sequence of gene activity during development. *Drosophila,* being amenable to extensive genetic manipulation, is particularly valuable for such studies [47,48]. The puffing patterns produced in response to MH in salivary glands of *Chironomus* are not as complex as those in *Drosophila* where six early and more than 100 late puffs have been recognized.

In the salivary gland chromosomes of *Drosophila,* two major sequences of puffing changes commence 6 h before puparium formation. All the puffs of the first sequence (larval) have regressed by 3–4 h after puparium formation; the second sequence (prepupal) begins some 5 h after puparium formation. For ease of precise hormonal manipulation, Ashburner and co-workers investigate control by MH of puffing in isolated salivary glands from intermoult third larval instar *D. melanogaster* cultured in a fully defined medium. Early and late puffs (not to be confused with larval and prepupal puffs) differ in many respects besides the timing of their response to ecdysone. In salivary glands of *Drosophila,* as in those of *Chironomus,* when protein synthesis is inhibited, normal early puff induction occurs, with failure of both their regression and late puff induction. Early puffs respond quantitatively over a wide concentration range (500 fold) of hormone, whereas the late puffs display an almost all-or-none reaction to the hormone. Finally, the activity of the early puffs is entirely dependent upon the continued presence of hormone, whereas the activity of late puffs does not require the presence of ecdysone, although *some* previous exposure of the glands to hormone is required for activity.

Ashburner's group have proposed a model to account for the control of the larval puffing sequence of *Drosophila,* with reference to one early and one late puff (Fig. 3.8) [47].

According to this hypothesis, the 'active' form of ecdysone is the complex with receptor protein. The early puff responds quantitatively as a function of the concentration of the complex and hence of the hormone itself, yielding a product P, which either is a protein or requires protein synthesis for its formation. P is presumed to repress the early puff itself and also to induce the late puff. If P is absent, as for example experimentally in the presence of cycloheximide, early puff regression and late

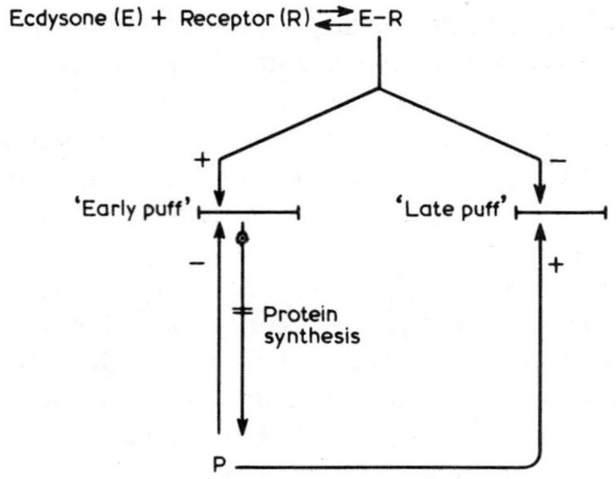

Fig. 3.8 Model for control by MH of an early and a late puff in larval salivary glands of *Drosophila*. Redrawn from [47].

puff induction are not observed. The ecdysone-receptor complex (ER) functions additionally to repress the late puff. It is envisaged that the concentration of ER is normally constant with time, but that of P increases, eventually becoming sufficient to displace ER from both early and late puff sites, resulting in the former puff's regression and the latter one's induction.

Whereas the larval puffing sequence is apparently controlled solely by an increase in MH concentration, regulation of the prepupal sequence is much more complex. Continuous exposure of larval glands to MH never causes the appearance of any puffs of the prepupal sequence. Apparently, some prepupal puffs require the absence of ecdysone for their activity, whereas induction of later ones requires a short hormone-free period before exposure to MH.

Conclusions drawn from studies of the puffing phenomenon should be applicable generally. No doubt, all active gene loci undergo structural modifications comparable with the microscopically discernible puffs of polytene chromosomes, though on a much smaller scale.

In order to explain the primary mode of action of MH, two main theories have been proposed: (i) a selective gene derepression hypothesis, and (ii) an ion hypothesis of gene activation.

3.5.2 Selective gene derepression hypothesis

The observation that injection of ecdysone into certain insects induces puffing in polytene chromosomes within 15 min, coupled with evidence that puffs represent sites of intensive transcription of the genetic information of the DNA into mRNA, led Karlson to formulate a possible biochemical mechanism of MH action [49]. According to this hypothesis, which makes use of the Jacob and Monod model of bacterial gene repression, the hormone penetrates to specific chromosomal sites and there interacts with the 'repressor molecule'. This results in derepression, so that mRNA synthesis begins at this site on the genome. Following transfer to the cytoplasm, the messenger participates in the biosynthesis of specific proteins, such as enzymes. The following predictions can be made from this hypothesis: (i) the hormone should exert its primary action in the nucleus, (ii) specific mRNA synthesis should be stimulated by the hormone, and (iii) formation of the respective enzyme protein should be induced. Considerable experimental support for this hypothesis has been obtained from extensive investigations of the process of sclerotization involved in puparium formation in the blowfly, *Calliphora erythrocephala* [50]. Sclerotization is due to the interaction of a quinone (produced by oxidation of N-acetyldopamine) with cuticular proteins. In late third instar larvae, production of MH causes a switch in tyrosine metabolism from transamination to formation of N-acetyldopamine (*see* Fig. 2.9). The rate-limiting enzyme in the synthesis of N-acetyldopamine is DOPA-decarboxylase, which is controlled by MH.

The predictions can be considered most conveniently in reverse order. The DOPA-decarboxylase activity rises almost simultaneously with the production of ecdysone (Fig. 3.3), and ecdysone injection into ligated abdomens of *Calliphora* produces increased DOPA decarboxylase activity some 7–10 h after injection. This rise in enzymic activity can be prevented by inhibitors of RNA synthesis, indicating that newly formed RNA is required for the process. Through a double-isotopic labelling technique and immunoprecipitation of the enzyme with specific antibody against DOPA decarboxylase it was demonstrated directly that the ecdysone-induced increase in enzyme activity is actually due to stimulation of enzyme synthesis *de novo* [51].

In line with the hypothesis, within 1 h after ecdysone injection into ligated larvae, stimulation of ^{32}P incorporation into the RNA of epidermis nuclei occurs. This labelled RNA fraction is apparently rich in mRNA as suggested by a number of criteria. A messenger RNA fraction isolated from epidermis of *Calliphora* stimulates protein synthesis, as measured by labelled leucine incorporation in a reconstituted system containing mouse liver ribosomal subunits. One of the proteins synthesized under the direction of insect mRNA has been identified as DOPA decarboxylase by immunoprecipitation and by co-migration with the pure enzyme on gel electrophoresis. The amount of translatable DOPA decarboxylase mRNA was 3–4 times higher in ecdysone-induced than non-induced larvae. Although this provides strong evidence for induction of synthesis of the specific mRNA by ecdysone, alternative interpretations of the results are still possible, e.g. 'activation' of a preformed 'inactive' mRNA by the hormone [52].

The presence of ecdysone in the nucleus was demonstrated by injection of labelled hormone into last instar *Calliphora* larvae, when a substantial amount of radioactivity was found in nuclei from epidermal tissue. Furthermore, when epidermis nuclei were incubated *in vitro* with ecdysone, they apparently responded to the hormone by increasing synthesis of RNA, probably mRNA since it had enhanced template activity. It is difficult to reconcile an important role of ecdysone-receptor proteins in the action of the hormone, with these studies. However, this work on isolated nuclei may warrant re-examination, since analogous stimulation of nuclear RNA synthesis by several mammalian hormones has been observed only when the animals were treated with hormone *in vivo.*

The main predictions made from the hypothetical scheme of transcriptional control by ecdysone are thus apparently well supported by the corresponding experiments. However, many lacunae exist, particularly regarding the precise nature of the interaction between the hormone (or its receptor complex) and the 'repressor' molecule. This hypothesis, together with the early supporting experiments raised considerable controversy and interest, and stimulated and formed a basis for much research on mammalian steroid hormones.

3.5.3 Ion hypothesis of gene activation

The ion hypothesis of gene activation proposed by Kroeger argues that ecdysone controls gene activity indirectly by controlling the electrolyte milieu of the nuclear sap. According to this theory, ecdysone exerts its primary action upon membranes, changing their selective permeability to small ions. This would lead to enhanced uptake of K^+ into the cell and nucleus possibly by stimulating the outward Na^+ pump. The increased intranuclear K^+ would then activate particular genes, possibly by interfering selectively with histone-DNA binding [53]. On the other hand, it is postulated that JH decreases the K^+/Na^+ ratio within the nucleus. Subsequently, a regulatory role for Mg^{2+} and Ca^{2+} has been proposed.

The ion hypothesis is supported by two main lines of evidence derived largely from experiments with salivary glands from *Chironomus spp.* [53,54]:

(i) Changes in genome activity (puffing) closely reminiscent of effects produced by ecdysone or JH can be elicited by artificial changes in the ionic concentrations of the medium surrounding the glands. For example, upon explantation of salivary glands into a series of high ionic strength solutions containing progressively increasing K^+/Na^+ ratios, the puff activity of certain loci changes from a typical larval pattern in high Na^+ to prepupal pattern in high K^+. Although it has not been possible to influence the puffing pattern in isolated chromosomes with the hormones, specific puffing can be induced in isolated chromosomes and nuclei by inorganic ions, particularly K^+ and Na^+, within seconds of application. All the chromosomal regions that become activated by JH action *in vivo* form a puff when isolated nuclei are incubated in Na^+-containing medium. Similarly, all the regions that specifically react to MH *in vivo*, respond when isolated nuclei are exposed to a K^+-rich medium. This perfect correlation argues against the criticism that ionic effects on puffing are less specific than the hormonal effects [55].

Preliminary electrophysiological studies do in fact suggest that MH increases the salivary gland nuclear K^+/Na^+ ratio, whereas JH apparently decreases this ratio. Injection of the cardiac glycoside, ouabain, into *Tenebrio* pupae produces pupal-adult intermediates, as does JH application. Since ouabain is believed to inactivate the mechanism for active extrusion of Na^+ from the cell, it would produce a lowered intracellular K^+/Na^+ ratio, consistent with a relationship between JH and high Na^+ concentration [1].

(ii) A redistribution of small ions across the cellular membranes due to ecdysone action should be reflected in the electrophysiological parameters of these membranes. In fact, the most rapid effect so far detected, 1 min after ecdysone application to excised *Ch. thummi* salivary glands, is a rise in the electrical potential difference across the nuclear and cell membranes. This correlates with the subsequent early induction of a particular puff, which is only formed when the potential difference rises above a certain threshold.

'Gene repression' is presumed to occur by binding of histone molecules via salt linkages to chromosomal DNA. According to Kroeger, the specific response of chromosomal loci to variations in the electrolyte milieu is based upon differences in sensitivity of the histone-DNA binding forces along the chromosome to such changes [56]. However, experimental support for this notion is lacking.

This ion theory of gene activation has been subject to much scepticism and controversy (*see* [24,57]). The experimental evidence, on the whole, does not rule out the possibility of non-specific effects, and the experiments *in vitro* used very high ecdysone concentrations (approx. 150 $\mu g\ ml^{-1}$). A critical appraisal of the principal evidence for the Kroeger ion hypothesis has led to the conclusion that ecdysone's effect on puffing is probably not mediated via the $[K^+]/[Na^+]$, although such a model cannot be excluded completely [58].

More recently, Kroeger has further modified his hypothesis and suggested a double action for ecdysone and/or ecdysone-receptor complex [57]. This involves a 'general' effect at the nucleus resulting in increased RNA synthesis, and control of ion flow at the cell membrane producing activation of specific genes (puffs) via changes in ionic milieu. In contrast to the original hypothesis, this modified theory would be consistent with many observations, such as the occurrence of ecdysone-binding proteins in nuclei of target tissue. A major question which remains unanswered concerns the exact mechanism of ecdysone action at the membranes to elicit changes in the intracellular electrolyte milieu.

3.5.4 Cyclic AMP

Besides its possible effect on membrane permeability, MH affects the membrane-bound enzyme, adenyl cyclase, which catalyses cyclic AMP formation from ATP. For example, incubation of *Drosophila* salivary glands *in vitro* with β-ecdysone produced a four-fold elevation in the cAMP content of the tissue [59]. However, cAMP did not induce ecdysone-specific puffs in such glands *in vitro*, but significantly increased the puff size (a measure of gene activity) of loci induced by β-ecdysone. Theophylline, which is an inhibitor of the enzyme (phosphodiesterase) responsible for breakdown of cAMP, produced a similar enhancement of the β-ecdysone effect, but produced no detectable effect upon hormone-specific gene activities.

Another system which has been used to investigate the possible intermediacy of cAMP in MH action is the stimulation of adult development by the hormone in diapausing pupae of *Hyalophora gloveri* [1]. β-Ecdysone stimulated adenyl cyclase in the pupal wing epidermis both *in vivo* and *in vitro*. Stimulation of adenyl cyclase occurred *in vitro* within minutes of hormone addition, and preceded increases in the rate of RNA and protein synthesis. However, injection of cAMP into debrained pupae did not initiate development, neither did it mimic all the initial effects (e.g. RNA synthesis) of β-ecdysone on wing epidermis *in vitro*. It may be significant that β-ecdysone also stimulates guanyl cyclase in this system.

Results from these and other studies suggest that β-ecdysone has effects independent of the activation of adenyl cyclase. Although this activation may be a non-specific effect, the rapidity of the response indicates that it could be an event close to the site of action of the hormone. The salivary gland chromosome experiments discussed previously suggest that cAMP acts synergistically with MH. Since there are examples of cAMP acting at the translational level, presumably by activation of a relatively stable mRNA, it is conceivable that MH acts indirectly, at least in part, upon the translation process.

3.6 Biochemical mode of action of juvenile hormones

A major effect of JH is to control sequential polymorphism i.e. it is involved in the switching on and off of successive batches of genetic information. Elucidation of the molecular mechanism of JH action is complicated by several conceptual and technical difficulties. It is not easy to disentangle the action of JH in insect metamorphosis, since the simultaneous action of MH to induce moulting is necessary for manifestation of the morphogenetic effects i.e. JH modifies the cell's response to MH, at least in pre-adult stages. Experiments *in vivo* may be subject to complications since JH may activate the prothoracic glands, thus elevating the ecdysone titre.

The question arises as to whether JH binds to a specific receptor in its target organs. By definition, a target organ accumulates hormone against a concentration gradient and degrades it more slowly than non-target organs do. Epidermal cells of *Tenebrio* pupae accumulate a labelled JH analogue, while at the same time it is degraded rapidly in other organs. Tentative evidence that the receptor may be a ribonucleoprotein is reminiscent of steroid hormone binding ribonucleoprotein particles of target organs in mammals [60].

JH-induced changes in puffing activity can be demonstrated in the presence of a high ecdysone titre (e.g. in prepupae), but not in its absence. However, the results of different groups of workers are somewhat conflicting. Lezzi and associates reported that application of JH to prepupae of *Chironomus tentans* elicited in salivary glands within 2–4 h, formation of a new puff and regression of Balbiani ring 1. Since ecdysone induced the opposite effects, they suggested that the two hormones may act antagonistically, possibly via a shift in K^+/Na^+ ratio [61]. The observation by Laufer's group that an apparently ecdysone-specific puff was not inhibited by JH, as well as their inability to demonstrate the previously reported antagonism between ecdysone and JH led them to suggest that JH acts in concert with, rather than antagonistically to, ecdysone [62]. According to this view each hormone stimulates or represses specific loci which are different from those affected by the other. It is possible that some of the discrepancies could be explained by differences in the administered dose of JH, since application of JH to pharate adults (i.e. prior to adult ecdysis) of *Ch. tentans* caused, within 1 h, the formation of a specific puff in the Malpighian tubule chromosomes, but the higher the dose the sooner the puff disappeared [63]. It remains to be established whether the JH-induced puffing changes occur with the rapidity demonstrated for ecdysone. The possibility exists that application of JH at the prepupal period may not produce typical effects, since the cells may already be committed to metamorphosis, and not be fully competent to respond to the hormone.

There is some evidence to suggest that JH may function at three levels of organization.

3.6.1 Transcriptional control

Administration of JH to *H. gloveri* pupae elicits the appearance of JH-esterases in the haemolymph some 6 h later. The fact that similar results are obtained in brainless pupae suggests that JH is not acting indirectly via stimulation of the prothoracic glands or the brain. Although inhibitor studies suggest that the appearance of these JH-esterases requires prior RNA and protein synthesis, and that the JH may be acting at the transcriptional level, the possibility of modification of pre-existing enzyme cannot be eliminated [21].

An extensively investigated action of JH is the induction of synthesis of vitellogenins (female-specific proteins) in the fat body of female adult insects [64]. The complete inhibition of synthesis of this protein by actinomycin D, together with other indirect evidence, suggests that JH controls transcription of specific mRNA [65]. The induction of both JH-esterase and vitellogenin synthesis probably constitutes a more 'positive' action of JH than its action in controlling metamorphosis, where it generally acts in conjunction with MH; the basic action of the hormone in these two types of processes may differ.

Although JH apparently does not affect developmental processes unless tissues have been exposed, previously or simultaneously, to MH there are numerous reports indicating some sort of antagonism between the action of the two hormones. For example, stimulation of RNA synthesis in isolated larval *Calliphora* fat body nuclei has been observed after 10 minutes exposure *in vitro* to either ecdysone or JH, but not in the presence of both hormones. Although these results may be interpreted as suggesting transcriptional control by both hormones, the possibility that at least part of the normal hormonal action might be via the cytoplasm cannot be completely excluded; it is difficult to prepare completely clean and physiologically normal nuclei. RNA-DNA hybridization studies demonstrated that new species of RNA are induced by the hormones, and led to the working hypothesis that some genes may be regulated by both hormones at the same time. These genes would be activated by one hormone alone, but inactivated by both hormones together [66]. In another example, an abnormally high concentration of JH inhibited reversibly ecdysone-induced evagination and stimulation of RNA synthesis in *Drosophila* imaginal discs cultured *in vitro.*

It was again proposed that JH may prevent the transcription of ecdysone-derepressed genes by acting as co-repressor [67].

A co-repressor role for JH is also envisaged in an intriguing detailed model proposed by Williams and Kafatos [68]. According to this hypothesis, the genome of each cell is divisible into three main gene-sets (larval, pupal, and adult), each being controlled by its own hypothetical 'master regulatory gene'. JH is proposed to be involved in the negative control (as a co-repressor) of the 'master regulatory genes' each of which codes for a specific sigma factor or RNA polymerase presumably having template specificity for the promotor sequence of all the genes in the corresponding gene-set. In essence, this hypothesis involves positive control of successive gene-sets by master regulatory genes which are themselves under JH-influenced negative control. This scheme calls for the demonstration of the occurrence of stage-specific RNA polymerases or sigma factors in insects; the occurrence of sigma factors in eukaryotic organisms generally has not yet been demonstrated.

With regard to possible transcriptional control by MH and JH it should be remembered that the ill-understood 'determinative' events which begin in the early embryo and continue on a decreasing scale during post-embryonic development dictate within individual cells those parts of the chromosomes which can and cannot be read under appropriate hormonal stimulus.

3.6.2 Translational control

In the Jacob and Monod model of microbial regulation of gene expression, it is prerequisite that the mRNA must be a short-lived intermediate in order to 'turn on' or 'turn off' synthesis of the final gene product (protein) by turning on or off the synthesis of mRNA. However, it is well established in many developing systems that many mRNAs are stable, and that there is usually a lag period between their synthesis and translation. Therefore, in cells containing stable mRNA, some of which is not translated immediately, there must be other mechanisms, acting at the post-transcriptional or translational level, that control gene expression.

A convenient system which Ilan and Ilan have used for study of JH action involves regulation of synthesis of adult cuticular protein during *Tenebrio* development. It is possible to regulate the development of pupa into adult by JH; topical application of JH on first-day pupae of *Tenebrio* causes a second pupal moult 8 days later. Pupal and adult cuticular protein can readily be distinguished by their different tyrosine content and amino acid sequence. Ilan and Ilan propose that in this system control of gene expression mediated by JH is at the translational level, and involves the appearance of new tRNA and its activating enzyme in the absence of hormone [69].

The mRNA for adult cuticular protein is present on the first day of pupation, and is translated maximally 7 days later. Polysomes from either day-1 or day-7 pupae synthesize the adult protein only in the presence of tRNA and amino acid activating enzyme from day-7 animals. When tRNA or amino acid activating enzyme from JH-treated animals are used to supplement the cell-free incorporating system, adult cuticular protein is not synthesized. These findings indicate that some of the polysomal mRNAs are translated selectively as a result of the appearance of rate-limiting aminoacyl-tRNA in day-7 pupae. The requirement for the amino acid activating enzyme as well as the tRNA could provide a double filter mechanism.

Since neither chain initiation nor chain termination of polypeptide occurs on the purified ribosomes, the translational effect is at the level of chain elongation. Of all the aminoacyl-tRNA synthetases only leucyl-tRNA synthetase activity apparently showed significant changes during *Tenebrio* metamorphosis and in response to treatment with JH. Day-7 pupae contain a new isoacceptor $tRNA^{Leu}$ which differs from the five isoacceptor $tRNA^{Leu}$ present in day-1 pupae. Evidence that the new isoacceptor $tRNA^{Leu}$ is involved with the switch in protein synthesis is indirect. The appearance of new leucyl specific tRNA during development has been reported for several developing systems besides insects. It is convenient to utilize $tRNA^{Leu}$ to regulate protein synthesis since the genetic code specifies the greatest number of codons (six) for leucine of all the amino acids, with the result that there should be six $tRNA^{Leu}$, which differ at least in the anticodon sequence of nucleotides.

Ilan and Ilan envisage the control operating as follows. If a group of mRNAs have a common code word for a rare species of tRNA, the appearance of this tRNA species will trigger the translation of these mRNAs. The degenerate nature of the code means that other messages, which do not share this particular code word, are independent of the hormone and can be translated. It is not known whether the appearance of the rate-limiting tRNA results from activation of pre-existing tRNA or from synthesis *de novo* by a direct effect of the hormone on DNA. If JH is concerned in selective transcription of tRNA, then the control of translation by the hormone is secondary.

Other studies by the same workers demonstrated the presence in *Tenebrio* of stage-specific initiation factors (ribosomal proteins), which can discriminate between mRNAs obtained from different developmental stages. These factors promote formation of the 80S ribosomal complex only with mRNA from the same stage of development, thus providing another mechanism for control of protein synthesis whereby ribosomes can select the mRNA to be translated. It remains to be demonstrated whether insect hormones are concerned in the induction of formation of such stage-specific initiation factors.

Much of the evidence for the translational control model proposed by Ilan's group for adult cuticular protein synthesis in *Tenebrio* has been criticized recently [70]. It has been claimed that detection of a new $tRNA^{Leu}$ during pupal development can be explained by the suboptimal conditions employed for amino-acylation of the leucine tRNAs, coupled with differential rates of charging of the leucine iso-acceptor tRNAs.

3.6.3 Membrane effects

The possible effect of JH upon membranes leading to increased nuclear Na^+ concentration has already been considered (p. 49). There is some other evidence that JH affects the properties of biological membranes [21]. For example, a depolarizing effect of the hormone upon the plasma membrane has been demonstrated.

Summary

The interaction between the effects of JH and MH at the molecular level poses a particular challenge in insect endocrinology. Each hormone does not necessarily have a single primary site of action, and may function by effects at several levels of organization. However, current evidence indicates that most steroid hormones apparently exert their main action by controlling transcription of specific genes, thus inducing protein synthesis. The mechanism of this control is not clear, and several plausible models have been proposed [57]. It is hardly surprising that elucidation of the mechanisms controlling gene expression in eukaryotic organisms has proved recalcitrant, since recent findings have led to a re-appraisal of the structural organization of chromatin itself. These findings will undoubtedly influence substantially future ideas regarding the molecular mechanism of gene switching by hormones.

References

[1] Gilbert, L. I. and King, D. S. (1973), in *The Physiology of Insecta, Second Edition*, (Ed.)

M. Rockstein, Vol. I, pp. 249–370, Academic Press, New York.
[2] Doane, W. W. (1973), in *Developmental Systems: Insects*, (Ed.) S. J. Counce and C. H. Waddington, Vol. 2, pp. 291–497, Academic Press, New York.
[3] Vedeckis, W. V., Bollenbacher, W. E. and Gilbert, L. I. (1976), *Molecular and Cellular Endocrinol.*, **5**, 81–88.
[4] Ashworth, J. M. (1973), *Cell Differentiation*, Chapman and Hall, London.
[5] Hadorn, E. (1965), *Brookhaven Symp. Biol.* **18**, 148–161.
[6] Hadorn, E. (1963), *Develop. Biol.*, **7**, 617–629.
[7] Ursprung, H. and Nöthiger, R. (Ed.), (1972), *The Biology of Imaginal Disks*, Springer-Verlag, Berlin.
[8] Gehring, W. J. and Nöthiger, R. (1973), in *Developmental Systems: Insects*, (Ed.) S. J. Counce and C. H. Waddington, Vol. 2, pp. 211–290, Academic Press, New York.
[9] Siegel, J. G. and Fristrom, J. W. (1974), *Develop. Biol.*, **41**, 314–330.
[10] Roberts, D. B. (1975), in *Current Topics in Developmental Biology*, (Ed.) A. A. Moscona and A. Monroy, Vol. 9, pp. 167–189, Academic Press, New York.
[11] Horn, D. H. S. (1971), in *Naturally Occurring Insecticides*, (Ed.) M. Jacobson and D. G. Crosby, pp. 333–459, Marcel Dekker, New York.
[12] Rees, H. H. (1971), in *Aspects of Terpenoid Chemistry and Biochemistry*, (Ed.) T. W. Goodwin, pp. 181–222, Academic Press, London.
[13] Shaaya, E. and Sekeris, C. E. (1965), *Gen. Comp. Endocrinol.*, **5**, 35–39.
[14] Thompson, M. J., Kaplanis, J. N., Robbins, W. E. and Svoboda, J. A. (1973), *Advan. Lipid Res.*, **11**, 219–265.
[15] Hanaoka, K. and Ohnishi, E. (1974), *J. Insect Physiol.*, **20**, 2375–2384.
[16] Legay, J-M., Calvez, B., Hirn, M. and De Reggi, M. L. (1976), *Nature*, **262**, 489–490.
[17] Hagedorn, H. H., O'Connor, J. D., Fuchs, M. S., Sage, B., Schlaeger, D. A. and Bohm, M. K. (1975), *Proc. natn. Acad. Sci. U.S.A.*, **72**, 3255–3259.
[18] Lagueux, M., Hirn, M., De Reggi, M. and Hoffmann, J. A. (1976), *C. R. Acad. Sc. Paris, Ser. D.*, **282**, 1187–1190.
[19] Svoboda, J. A., Kaplanis, J. N., Robbins, W. E. and Thompson, M. J. (1975), *Ann. Rev. Entomol.*, **20**, 205–220.
[20] Galbraith, M. N., Horn, D. H. S., Middleton, E. J., Thomson, J. A. and Wilkie, J. S. (1975), *J. Insect Physiol.*, **21**, 23–32.
[21] Gilbert, L. I. (1974), in *Recent Progress in Hormone Research*, **30**, 347–390.
[22] Hsiao, T. H., Hsiao, C. and de Wilde, J. (1975), *Nature*, **255**, 727–728. See also references cited therein.
[23] Hoffmann, J. A. and Koolman, J. (1974), *J. Insect Physiol.*, **20**, 1593–1601.
[24] Wyatt, G. R. (1972), in *Biochemical Actions of Hormones*, (Ed.) G. Litwack, Vol. II, pp. 385–490, Academic Press, New York.
[25] Locke, M. (1974), in *The Physiology of Insecta, Second Edition*, (Ed.) M. Rockstein, Vol. VI, pp. 123–213, Academic Press, New York.
[26] Willis, J. H. (1974), *Ann. Rev. Entomol.*, **19**, 97–115.
[27] Bergtrom, G. and Oberlander, H. (1975), *J. Insect Physiol.*, **21**, 39–52.
[28] Milner, M. J. and Sang, J. H. (1974), *Cell*, **3**, 141–143.
[29] Borst, D. W. Bollenbacher, W. E., O'Connor, J. D., King, D. S., and Fristrom, J. W. (1974), *Develop. Biol.*, **39**, 308–316.
[30] Bodnaryk, R. P. (1975), *Life Sci.*, **16**, 1411–1416.
[31] Röller, H. and Dahm, K. H. (1968), in *Recent Progress in Hormone Research*, **24**, 651–680.
[32] Meyer, A. S., Hanzmann, E., Schneiderman, H. A., Gilbert, L. I. and Boyette, M. (1970), *Arch. Biochem. Biophys.*, **137**, 190–213.
[33] Judy, K. J., Schooley, D. A., Dunham, L. L., Hall, M. S., Bergot, B. J. and Siddall, J. B. (1973), *Proc. natn. Acad. Sci. U.S.A.*, **70**, 1509–1513.
[34] Lanzrein, B., Hashimoto, M., Parmakovich, V., Nakanishi, K., Wilhelm, R. and Lüscher, M. (1975), *Life Sci.*, **16**, 1271–1284.
[35] Schooley, D. A. Judy, K. J., Bergot, B. J., Hall, M. S. and Siddall, J. B. (1973), *Proc. natn. Acad. Sci. U.S.A.*, **70**, 2921–2925.
[36] Peter, M. G. and Dahm, K. H. (1975), *Helv. Chim. Acta*, **58**, 1037–1048.
[37] Slade, M. and Wilkinson, C. F. (1974), *Comp. Biochem. Physiol.*, **49B**, 99–103.
[38] Kramer, K. J., Sanburg, L. L., Kézdy, F. J. and Law, J. H. (1974), *Proc. Nat. Acad. Sci. U.S.A.*, **71**, 493–497.
[39] Sanburg, L. L., Kramer, K. J., Kézdy, F. J., Law J. H. and Oberlander, H. (1975), *Nature*, **253**, 266–267.
[40] Truman, J. W., Riddiford, L. M. and Safranek, L. (1974), *Developmental Biol.*, **39**, 247–262.

[41] Sláma, K. (1975), *J. Insect Physiol.*, **21**, 921–955.
[42] Nijhout, H. F. and Williams, C. M. (1974), *J. exp. Biol.*, **61**, 493–501.
[43] Solomon, K. R. and Walker, W. F. (1974), *Science*, **185**, 461–462.
[44] Ashburner, M. (1970), *Adv. Insect Physiol.*, 7, 1–95.
[45] Beerman, W. (Ed.), (1972), *Developmental Studies on Giant Chromosomes*, Springer-Verlag, Heidelberg.
[46] Lefevre, G. Jr. (1974), *Ann. Rev. Genet.* **8**, 51–62.
[47] Ashburner, M. and Richards, G. (1976), in *Insect Development*, (Ed.) P. A. Lawrence, pp. 203–225. Blackwells, Oxford.
[48] Fristrom, J. W. and Yund, M. A. (1973-Sept.), *C.R.C. Critical Reviews in Biochemistry*, pp. 537–570.
[49] Karlson, P. (1963), *Perspect. Biol. Med.*, **6**, 203–214.
[50] Karlson, P. and Sekeris, C. E. (1966), in *Rec. Progr. Hormone Res.*, **22**, 473–502.
[51] Fragoulis, E. G. and Sekeris, C. E. (1975), *Biochem. J.*, **146**, 121–126.
[52] Fragoulis, E. G. and Sekeris, C. E. (1975), *Eur. J. Biochem.*, **51**, 305–316.
[53] Kroeger, H. (1968), in *Metamorphosis – a Problem in Developmental Biology*, (Ed.) W. Etkin and L. I. Gilbert, pp. 185–219, North-Holland, Amsterdam.
[54] See chapter by Lezzi, M. and Robert, M. in [45].
[55] Lezzi, M. and Frigg, M. (1971), *Bull. Soc. Entomol. Suisse*, **44**, 163–170.
[56] Kroeger, H. (1971), *Endocrinol. Exp.*, **5**, 108–113.
[57] Karlson, P., Doenecke, D. and Sekeris, C. E. (1975), in *Comprehensive Biochemistry*, (Ed.) M. Florkin and E. H. Stotz, Vol. 25, pp. 1–63, Elsevier, Amsterdam.
[58] Ashburner, M. and Cherbas, P. (1976), *Molecular and Cellular Endocrinol.*, **5**, 89–107.
[59] Leenders, H. J., Wullems, G. J. and Berendes, H. D. (1970), *Exptl. Cell Res.*, **63**, 159–164.
[60] Schmialek, P. (1973), *Nature*, **245**, 267.
[61] Lezzi, M. and Gilbert, L. I. (1969), *Proc. natn. Acad. Sci. U.S.A.*, **64**, 498–503.
[62] Laufer, H. and Calvert, J. P. (1972), *Gen. Comp. Endocrinol. Suppl.*, **3**, 137–148.
[63] Holderegger, C. and Lezzi, M. (1972), *J. Insect Physiol.*, **18**, 2237–2250.
[64] Engelmann, F. (1974), *Amer. Zool.*, **14**, 1195–1206.
[65] Engelmann, F. (1972), *Gen. Comp. Endocrinol. Suppl. 3*, 168–173.
[66] Congote, L. F., Sekeris, C. E. and Karlson, P. (1970), *Z. Naturforsch.*, **25b**, 279–284.
[67] Chihara, C. J. and Fristrom, J. W. (1973), *Develop. Biol.*, **35**, 36–46.
[68] Williams, C. M. and Kafatos, F. C. (1971), *Bull. Soc. Entomol. Suisse*, **44**, 151–162.
[69] Ilan, J. and Ilan, J. (1975), in *Current Topics in Developmental Biology*, (Ed.) A. A. Moscona and A. Monroy, Vol. 9, pp. 89–136, Academic Press, New York.
[70] Lassam, N. J., Lerer, H. and White, B. N. (1975), *Nature*, **256**, 734–735.

Further reading

General review articles have been cited e.g. [1,2,24].

Wigglesworth, V. B. (1970), *Insect Hormones*, Oliver and Boyd, Edinburgh.
'A broadly based introductory text on insect hormones, concentrating primarily upon physiological aspects.'

Highnam, K. C. and Hill, L. (1976), *The Comparative Endocrinology of the Invertebrates*, second edition, Edward Arnold, London
'An introductory book on physiological aspects of invertebrate endocrinology with emphasis on insects. The importance of neurosecretory hormones in coordinating the development and reproduction of animals with their changing environment is stressed.'

Novák, V. (1975), *Insect Hormones*, 2nd English Edition, Chapman and Hall, London.
'A comprehensive advanced text covering primarily the physiology of insect hormones.'

Sláma, K., Romaňuk, M. and Šorm, F. (1974), *Insect Hormones and Bioanalogues*, Springer-Verlag, Wien.
'A comprehensive advanced text covering the physiology and biochemistry of insect developmental hormones.'

The proceedings of several symposia on insect developmental hormones have been published:
Menn, J. J. and Beroza, M., (Eds.) (1972), *Insect Juvenile Hormones – Chemistry and Action*, Academic Press, New York.
Gilbert, L. I., (Ed.) (1976), *The Juvenile Hormones*, Plenum Press, New York.
Burdette, W. J., (Ed.) (1974), *Invertebrate Endocrinology and Hormonal Heterophylly*, Springer-Verlag, Berlin.

'This book deals with brain hormone, ecdysones and juvenile hormones.'

Review: Gilbert, L. I., Goodman, W. and Bollenbacher, W. E. (1977 in press) in *MTP International Review of Science*, Biochemistry Series II, Vol. 4, T. W. Goodwin, (Ed.) Butterworths, London.
'The biochemistry of insect lipids is reviewed, with emphasis on the occurrence, biosynthesis and metabolism of ecdysones and juvenile hormones.'

4 Insects and the external environment: pheromones and defensive substances

The interactions of insects with their external environment may involve many phenomena e.g. bioluminescence, vision [1], pheromone communication, defence, and the relationship between plants and insects including control of feeding behaviour by chemical substances [2,3]. In this chapter we shall consider briefly pheromones and defensive substances. The incredible success of insects on this planet is due not only to their various adaptations of metabolism and developmental processes, but also to their ability to communicate with other members of the same species and to defend themselves against predators. Chemical signals constitute an important part of this communication, which may be associated with reproduction, protection and acquisition of food. The term 'pheromone' is applied to chemical compounds which enable members of the same species to communicate with each other.

4.1 Pheromones

The very nature of pheromone communication requires that airborne pheromones be volatile. Most insect pheromones are of relatively low molecular weight, and many are derivatives of terpenes and fatty acids. Pheromone receptors are located mainly in insect antennae, but the mechanisms whereby small molecules interact with some component in a receptor cell in such a way as to cause the cell to respond are uncertain.

Pheromones can be divided conveniently into two general types, releasers and primers. Releaser pheromones initiate immediate behavioural responses in insects upon reception, whereas primer pheromones produce physiological changes in an animal which ultimately result in a behavioural response. Releasers act principally through the nervous system, while primer action usually involves the endocrine system as well. Releaser pheromones may cause sexual attraction, alarm behaviour, recruitment, and aggregation [4].

4.1.1 Sex pheromones

The sex attractants are usually released by the female insect to lure the male from a distance and frequently to excite the male sexually. Sex pheromones released by males primarily function in sexually exciting the female. As an example, 12 mg of the sex attractant of the silkmoth, *Bombyx mori*, was originally isolated from half a million virgin females, and identified as (10*E*, 12*Z*)-hexadecadien-l-ol (bombykol; Fig. 4.1). Insects are highly sensitive and specific for the sex pheromone of their species, and require specific geometric and optical isomers. For example, *B. mori* males will apparently respond to air streams containing as little as 200 molecules of bombykol/cm^3, but the biological activity of bombykol is almost eliminated by a single change in the stereochemistry of one of the olefinic bonds.

Sex attraction processes may be complex since many sex attractants consist of several components, the characteristic nature of such mixtures undoubtedly eliminating cross-mating

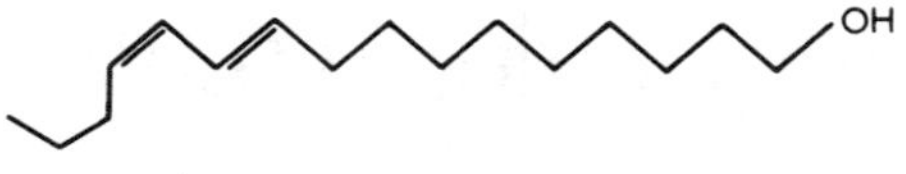

Fig. 4.1 (10*E*, 12*Z*) hexadecadien-1-ol.

of species. Furthermore, secondary chemicals which enhance or negate the potency of sex pheromones without exhibiting any noticeable behavioural effect themselves, are commonly present.

4.1.2 Alarm pheromones

Alarm pheromones are employed principally by social insects to warn other members in the colony of impending danger. In combat among social insects, the contents of the mandibular glands are generally discharged onto the enemy, which then becomes labelled as an aggressor. By diffusion of the pheromone vapours through the air, alarm is communicated to other members of the society. The alarm substances of ants are primarily terpenes and non-terpenoid hydrocarbons. Insects show far less sensitivity and chemospecificity for alarm pheromones than for sex pheromones. Alarm selectivity is apparently based more upon volatility than on unique structural characteristics. For example, in the ant, *Iridomyrmex pruinosus*, the alarm response to the natural pheromone, 2-heptanone, is mimicked by many other acyclic ketones of comparable volatility.

4.1.3 Recruitment pheromones

The recruitment (trail-marking) pheromones function as navigational aids in directing other members of the colony to a distant location. Recruitment by chemical communication is widespread in social insects (bees, termites, and ants). Most frequently, the trails are laid by foraging workers as they return from a food source. For example, in the honeybee, *Apis mellifera*, the constituents of the Nassanoff gland contents (Fig. 4.2) apparently mark the nest and rich food sources.

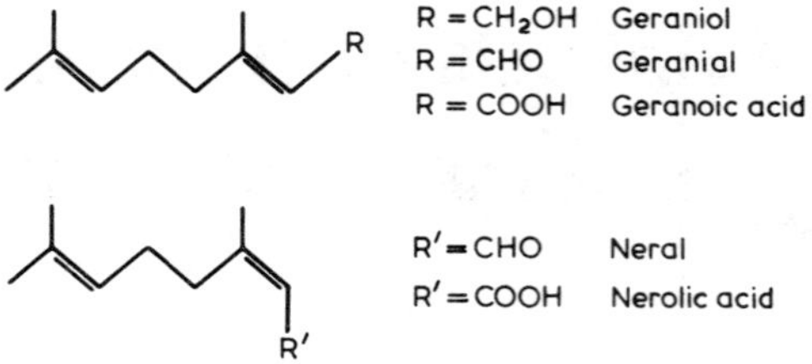

Fig. 4.2 Nassanoff gland contents of *Apis mellifera.*

4.1.4 Aggregation pheromones

Aggregation pheromones are used for aggregation to a feeding or mating site, and occur mainly among Coleoptera. In such insects, feeding on the host triggers the production and release of pheromone, frequently causing mass infestation at the particular site.

4.1.5 Primer pheromones

Primer pheromones activate chemoreceptors in such a manner as to change the physiology of the receiving organism, probably in most cases via the endocrine system, so that the animal shows a different response pattern in the future. In some cases a single pheromone can have a variety of functions. For example, (2*E*)-9-oxo-2-decenoic acid released by the virgin queen bee serves as a sex pheromone to lure drones to the receptive female during mating flights. After the newly mated queen has returned to her hive she produces a mixture of (2*E*)-9-oxo-2-decenoic acid and (2*E*)-9-hydroxy-2-decenoic acid (Fig. 4.3), which fulfils primer functions by inhibiting the development of another queen and also the development of workers' ovaries.

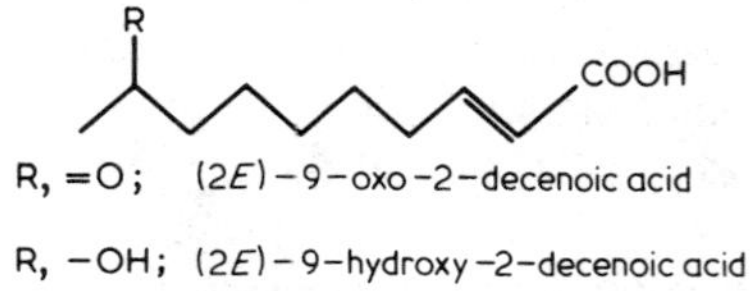

Fig. 4.3 Chemical structures.

4.2 Defensive chemistry

The insect defence substances, although of diverse chemical classes, are generally fairly simple compounds stored in high concentrations usually in the defensive organs, and rarely truly toxic (in the human sense) [3,5]. The chemicals may be effective for a number of reasons, e.g. they may be distasteful, malodorous, topical irritants, or systemic poisons. The question as to how a living system can produce and store potent toxicants without poisoning itself in the process is pertinent. This is frequently achieved by spatial separation of the final steps in the biosynthesis of the toxic end-product, a phenomenon which is well illustrated in the bombardier beetles (*Brachynus* spp.) [6]. Only when danger threatens do these beetles produce toxic quinones and explode them in the direction of the attacker as a hot poisonous cloud. The insects store in their collecting bladder not the unstable quinones, which would react with protein, but hydroquinones in a hydrogen peroxide solution. This hydrogen peroxide may frequently attain a concentration as high as 28%. During defence a hot toxic gas consisting of benzoquinone, toluquinone, unreacted hydroquinone, and hydrogen peroxide are audibly emitted by oxygen under pressure. The explosive reaction occurs in explosion chambers, which are thick chitin capsules, into which enzyme glands open. Two reactions occur in the chambers:

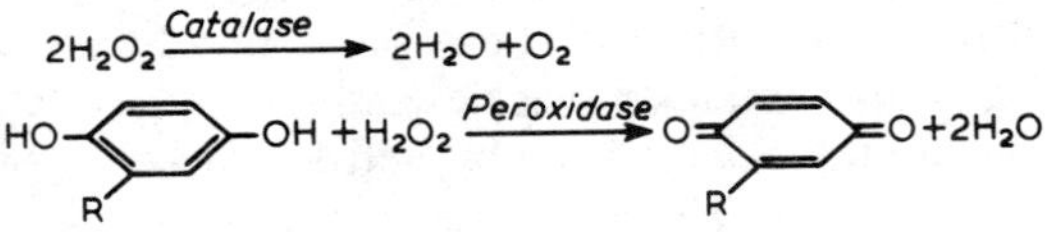

The *Brachynus* catalase is unusual in that its temperature optimum is 70°–80°, and it even shows considerable activity after being boiled for a short time. The peroxidase is distinctive in that it is appreciably more stable to hydrogen peroxide than are other peroxidases. This catalase system is particularly suited for an explosive reaction, since it is characteristic of catalases that their activity increases without limit with increasing substrate concentration. The characteristics of the enzyme, the high substrate (H_2O_2) concentration, and the unusually high reaction temperature in the chitin capsule, result in such rapid evolution of gas that an explosion occurs. Therefore, the denaturing effect of H_2O_2 on the enzymes is hardly significant. This defensive system offers a prime example of the extensive adaptation by insects of common biochemical reactions to serve a specialized function.

References

[1] Rockstein, M. (Ed.) (1974), *The Physiology of Insecta* 2nd Edition, Vol. 2, Academic Press, New York.

[2] Wood, D. L., Silverstein, R. M., Nakajima, M., (Eds.) (1970), *Control of Insect Behaviour by Natural Products,* Academic Press, New York.

[3] Sondheimer, E. and Simeone, J. B. (Ed.) (1970), *Chemical Ecology*, Academic Press, New York.

[4] Jacobson, M. (1974), in *The Physiology of Insecta* 2nd Edition, (Ed.) M. Rockstein, Vol. 3, pp. 229–276, Academic Press, New York.

[5] Weatherston, J. and Percy, J. E. (1970), in *Chemicals Controlling Insect Behavior* (Ed.) M. Beroza, pp. 95–144, Academic Press, New York.

[6] Schildknecht, H. (1971), *Endeavour*, **30**, 136–141.

5 New approaches to insect control

The majority of the broad-spectrum chemical insecticides now used act on the insect nervous system. Such insecticides lack selectivity, and kill helpful insects in addition to harmful ones, and may also be toxic to birds and mammals. Of particular concern is the fact that their use has caused the emergence of insecticide-resistant insect strains. Current attempts to overcome these difficulties in the development of novel insecticides involve exploiting aspects of insect biochemistry and behaviour for which there is no counterpart in vertebrates. Particular emphasis is placed upon exploiting pheromone communication and the hormonal regulation of moulting and metamorphosis.

Literally hundreds of JH-analogues have been synthesized and tested biologically, and in fact, a few compounds have been introduced commercially for field use [1]. The principal lethal effect of JH-active compounds, which readily penetrate the insect cuticle, is to interfere with the larval-pupal, pupal-adult, and nymphal-adult moults, as well as normal embryonic development. However, the majority of current analogues do not block the development of the growing larvae or nymphs, which in many phytophagous insects is the most destructive stage. In addition, developmental periods which are susceptible to JH compounds are often of short duration. It may be possible to circumvent these difficulties, since certain amines, which are structurally related to compounds with JH activity in insects, inhibit development at all larval stages [2]. The mode of action of these amines is uncertain. Although the specificity of JH analogues is typically less than that of most pheromones, by careful choice of analogues, it is possible to attack selectively certain insect pests.

Developmental hormones and pheromones in insects regulate physiological processes and behaviour essential for survival of the species. Therefore, a strong argument for the development of such compounds or closely related chemicals for insect control is that insects are less likely to develop resistance to them than to conventional insecticides. However, there is some indication that certain insecticide-resistant insects may also show some degree of cross-resistance to JH analogues.

The ecdysone-mediated processes constitute a prime target area for development of hormonal insecticides, since every nymphal or larval moult would be vulnerable to chemicals disrupting these processes. The moulting hormones themselves have disadvantages for use in interfering with these processes in that they are complex molecules and do not readily penetrate the cuticle, being active primarily by ingestion. These disadvantages can only be partly overcome by synthesis of simpler, more active moulting hormone compounds.

In addition to the development of compounds with insect hormone activity, a prime goal in the development of novel insecticides is the investigation of hormone antagonists or antihormones, and compounds which interfere with insect hormone biosynthesis and metabolism. An excellent example of compounds which interfere with moulting hormone biosynthesis and metabolism is furnished by numerous potent 25-azasteroids, which inhibit insect growth, moulting and metamorphosis [3]. Such azasteroids exhibit a great deal of species selectivity. Reports of inhibitors of JH biosynthesis are also beginning to appear.

A major stimulus to research on pheromones has been the prospect that they might lead to increased selectivity in insect control, and also diminish the disruptive effects of insecticides on the environment. The sex attractants are the ones most frequently considered for insect control programmes, although recruitment or aggregation pheromones have also been used [4]. Three principal approaches to the use of pheromones in pest control are currently being explored:

(i) The early detection and monitoring of infestation by a pest, employing traps baited with attractant, allows more judicious application of broad spectrum insecticides.
(ii) The mass trapping technique employs traps baited with sex pheromone, which lures males to the traps rather than to the females.
(iii) The 'male confusion' technique depends upon permeation of the air with female sex pheromone, so that the male, surrounded with the attractant becomes confused and unable to locate the female.

A new class of insecticides under development having a 1-(2,6-dichlorobenzoyl)-3-arylurea structure apparently act by blocking synthesis of endocuticular chitin, and are, thus, presumably specific for those invertebrates requiring this polymer. These compounds are larvicides, death of the insect being due to rupturing of the new cuticle [5].

Natural products possessing insecticidal activity occur widely, and have been investigated in such diverse organisms as plants (e.g. pyrethrins), arthropods (various toxins and venoms), bacteria (e.g. *Bacillus thuringiensis* exotoxins), fungi, and viruses [6,7]. Although most of the naturally occurring insecticidal compounds will probably not be used as such for insect control, they often represent truly novel structural types and can provide leads for the design of synthetic compounds having increased potency and specificity.

Finally, mention should be made of investigations of truly biological control methods. These include the possible use of insect predators and parasites, and natural pathogenic bacteria, viruses, fungi, protozoa, and nematodes [7], as well as genetic methods [8]. In fact, both bacterial and viral control of insects is a reality in the field.

References

[1] Staal, G. B. (1975), *Ann. Rev. Entomol.*, **20**, 417–460.

[2] Robbins, W. E., Thompson, M. J., Svoboda, J. A., Shortino, T. J., Cohen, C. F., Dutky, S. R. and Duncan, O. J. (1975), *Lipids,* **10**, 353–359.

[3] Thompson, M. J., Serban, N. N., Robbins, W. E., Svoboda, J. A., Shortino, T. J., Dutky, S. R. and Cohen, C. F. (1975), *Lipids,* **10**, 615–622.

[4] Beroza, M. (1972), *Toxicological and Environmental Chemistry Reviews*, **1**, 109–134.

[5] Post, L. C., de Jong, B. J. and Vincent, W. R. (1974), *Pesticide Biochem. and Physiol.*, **4**, 473–483.

[6] Jacobson, M. and Crosby, D. G. (Ed.) (1971), *Naturally Occurring Insecticides*, Dekker, New York.

[7] Bulla, L. A., Jr. (Ed.) (1973), *Ann. New York Acad. Sci.*, **217**.

[8] Whitten, M. J. and Foster, G. G. (1975), *Ann. Rev. Entomol.*, **20**, 461–476.

Further Reading

Jacobson, M., (Ed.) (1975), *Insecticides of the Future*, Dekker, New York.

Index